W9-BSS-390

THE TELESCOPE

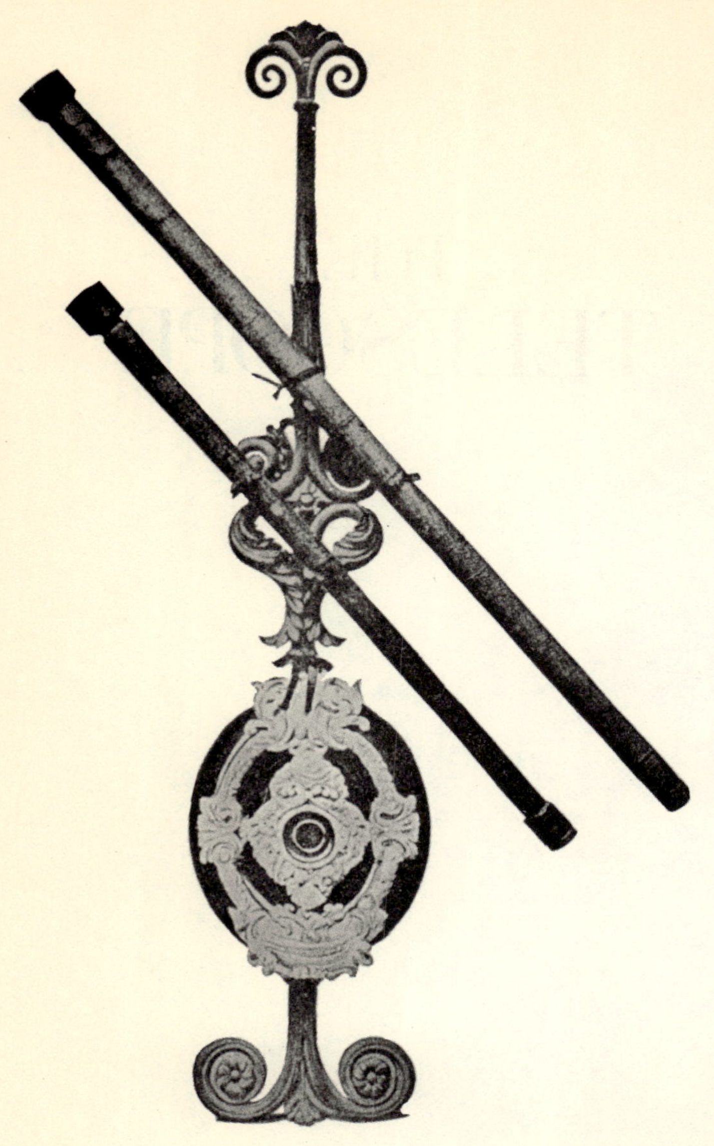

Galileo's Telescopes. *(Frontispiece)*

(Bull. de la Soc. Astron. de France.)

THE TELESCOPE

Louis Bell, Ph.D.

With a New Introduction by

PROF. JAY M. PASACHOFF
Director, Hopkins Observatory
Williams College

Dover Publications, Inc.
New York

Published in Canada by General Publishing Company, Ltd., 30 Lesmill Road, Don Mills, Toronto, Ontario.
Published in the United Kingdom by Constable and Company, Ltd., 10 Orange Street, London WC2H 7EG.

This Dover edition, first published in 1981, is an unabridged republication of the work originally published in 1922 by McGraw-Hill Book Co., Inc., New York. A new introduction has been written by Prof. Jay M. Pasachoff especially for this edition.

International Standard Book Number: 0-486-24151-3
Library of Congress Catalog Card Number: 80-68449

Manufactured in the United States of America
Dover Publications, Inc.
180 Varick Street
New York, N.Y. 10014

INTRODUCTION TO THE DOVER EDITION

Optical astronomy is holding its own in this era of telescopes in space. After a relative hiatus in the building of large telescopes, the 1970s brought a spate of construction of new large telescopes. At the same time, the introduction of modern electronic instrumentation and new film techniques made both new and old telescopes much more efficient. Further, amateur astronomy adopted other types of telescope designs besides the classical Newtonian.

Louis Bell's *The Telescope*, printed in 1922, is surprisingly timely and up to date. Perhaps we are looking back to the wisdom of the past, as the paraboloidal mirror of the 1940's 5-m Hale telescope on Palomar Mountain has given way to the Ritchey-Chrétien design by which most large telescopes are now constructed. Bell's discussions are wide-ranging and fundamentally sound, so that the lessons he teaches us about basic telescope design still remain valid.

As Bell writes in his chapter on "The Properties of Objectives and Mirrors" (p. 94-5), "the weak point of the parabolic mirror is in dealing with rays coming in parallel but oblique to the axis." Ira S. Bowen, a former Director of the Palomar Observatory, observed, (in his George Darwin Lecture to the Royal Astronomical Society, reprinted in their *Quarterly Journal* in 1967) that even with a corrector lens near the film, the Palomar telescope gives "moderately good definition over a field 15′ in diameter. However, this small field which requires thirty-two plates to make a mosaic covering one square degree is in my opinion the one feature in which the 200-inch telescope has proved disappointing."

Bell suggests a way to improve the situation, but his book predates the development around 1930 of a wide-field camera by the German optician Bernhard Schmidt. Another approach to the problem is the Ritchey-Chrétien design, which was suggested by the French astronomer M. H. Chrétien in 1922. The American optician Ritchey was the first to perfect the difficult aspheric surfaces that were required for this version of a Cassegrain system. The curvature of the objective is a hyperboloid, a few percent deeper than that of a paraboloid, and the aberrations introduced are removed by providing the secondary with its own complicated shape. Spherical aberration and coma are eliminated, leaving only astigmatism and field curvature. The field in good focus at the Cassegrain focus and, with a correcting lens near the film, at the prime focus, can be relatively large. The field of view of such new Ritchey-Chrétien telescopes as the Kitt Peak and Cerro Tololo 4-m reflectors is almost 1°, substantially larger than the 2 minutes of arc (1 cm across) available coma-free at the prime focus of the Palomar paraboloid and larger than the 15 minutes of arc field available with the Palomar prime-focus corrector. The advantages this provides allow, for example, better study of extended objects. The wide field also allows improved guiding of all faint objects when observing in directions away from the Milky Way, where suitable guide stars of 14th or 15th magnitude are separated from each other at an average distance of 5 to 10 minutes of arc.

Compound instruments, in which a thin lens provides correction for aberrations that would otherwise be introduced by a spherical objective mirror, are now quite common among telescopes for amateur astronomers. The Schmidt-Cassegrain has a thin aspheric correcting plate and a spherical main mirror. The convex secondary is attached to the back of the corrector. The Maksutov-Cassegrain uses a thick-miniscus correcting plate and a spherical main mirror. Its secondary mirror is a reflecting spot directly on the back of the corrector. Such telescopes, in sizes from 8 cm (3½ in) to 35 cm (14 in), are widely available and are in general use. Because their optical paths are folded, they are compact and easy to transport. Wide-field Schmidt cameras of this amateur grade and price are also available. The high-quality photographs taken with these commercial telescopes testify to the high optical quality of the instruments.

Prototypes of the current generation of large telescopes are the Ritchey-Chrétien designs of the 4-m reflectors at the Kitt Peak National Observatory near Tucson, Arizona, and the Cerro Tololo Inter-American Observatory in Chile. The telescopes were opened in 1974 and 1975, respectively, and through their visitors' programs are at the center of research for many American astronomers. Indeed, when observing time is granted at one of the telescopes, funds to cover all but the first $200 of travel expenses are automatically granted as well so that no one is penalized because of geographic location. Both Kitt Peak and Cerro Tololo are funded by the National Science Foundation, and are run by Associated Universities for Research in Astronomy (AURA), a consortium of universities.

The dependence of professional astronomers on a few observing centers, each with many telescopes, is now well established. Sixteen telescopes are now located on Kitt Peak, most newly constructed there and some moved there from other sites. Over half a dozen telescopes stand on Cerro Tololo. Another dozen telescopes are located on another Chilean peak, La Silla. Here the European Southern Observatory, operated by a consortium of European universities and observatories, has located both telescopes that belong to the countries individually and telescopes that are jointly owned. A 3.6-m reflector is the largest. A 1-m Schmidt camera has been constructed with plate scale to match the Palomar Schmidt. Nearby, the Carnegie Institution of Washington has located both a new 2.5-m and a new Schmidt camera on Cerro las Campañas. The site is operated together with its parent site in California as the Mt. Wilson and las Campañas Observatories.

The location of so many telescopes in Chile comes, of course, from the need to observe the numerous interesting objects in the southern sky, coupled with the particular advantages of the Chilean sites for astronomical observing. The coastal range west of the Andes has both clear skies and exceptional seeing.

Australia has long been as astronomical center. The 3.9-m Anglo-Australian Telescope (AAT), a joint venture of the United Kingdom and Australia, is located on Siding Spring Mountain in eastern Australia. The 1.2-m Schmidt there is being used in conjunction with the ESO Schmidt in Chile to map the entire southern sky as an extension of the Palomar

Schmidt's survey of the entire sky that it could see. The U.K. Schmidt is observing in the blue; plates are returned to the Royal Observatory in Edinburgh for examination. The ESO Schmidt is carrying out the red part of the survey. Both are using new film emulsions that are far superior to those in use at the time of the original Palomar survey. At the same time, the Palomar Schmidt is undertaking an infrared survey of the northern sky.

The latest site under major development is Mauna Kea, a 13,800 ft (4,200 m) extinct volcano on the island of Hawaii in the state of Hawaii. One of the major factors leading to the extensive use of this site is the low water-vapor content overhead, which makes it especially suitable for infrared observations. The 3.8-m (150-in) United Kingdom Infrared Telescope, the 3.6-m (140-in) Canada-France-Hawaii Telescope, and the 3-m (120-in) NASA-University of Hawaii infrared telescope all opened here in 1979. The UKIRT is the world's largest telescope especially designed for infrared observations. One novel feature is its use of a thin objective mirror, only 80 tons in weight instead of the more normal 300 to 400 tons for a mirror that size. The experiment seems to be a success, in that the images achieved are comparable with those obtained with mirrors of conventional thickness.

The infrared telescopes on Hawaii are especially designed to minimize the amount of infrared radiation from the telescope structure entering the detectors. The 2.3-m (92-in) Wyoming Infrared Telescope on Jelm Mountain near Laramie has also been specially optimized for infrared work. For example, the size of the Cassegrain opening is minimized to limit the amount of radiation that can pass through it and be reflected into the detector. The amount of telescope structure that can be viewed by the detectors is minimized, and suitable parts are cooled. Secondary mirrors that "chop" back and forth several times a second allow the object being observed to be compared with blank sky and are standard on infrared telescopes.

The largest optical telescope in the world is the 6-m telescope at the Soviet Special Astrophysical Observatory on Mount Pastukhov in the Caucasus. After years of construction, it went into service in 1976. Working out the bugs apparently has been a time-consuming task, but good data were available by the time of the 1979 International Astronomical Union meeting. The original mirror, made of Pyrex, has been replaced with another Pyrex mirror, and a further change to a ceramic mirror is planned for about 1983 when the material becomes available in the Soviet Union.

The strangest current design for a telescope belongs to the Multiple Mirror Telescope on Mount Hopkins, south of Tucson, Arizona. The MMT, now the third largest telescope in the world, opened in 1979. It contains six 1.8-m (72-in) mirrors mounted together on a tracking structure. The mirrors are aligned by a feedback loop that keeps track of laser beams which follow through the six independent systems. This method provides alignment of the telescope beams to about 1 arc sec, comparable to the limit that seeing imposes in any case. Mount Hopkins was chosen in part because it is 2,606 m (8,550 ft) high, allowing infrared work. The MMT is a joint venture of the Smithsonian Astrophysical Observatory and the University of Arizona.

Both the Soviet 6-m telescope and the MMT are on altazimuth mounts. The existence of computer controls now compensates for the fact that constantly changing rates of motion on two axes must be made. One advantage of the altazimuth mount is that the structure that houses the telescope—a conventional dome for the Soviet telescope and a rotating rectangular structure for the MMT—can be made much smaller than it need be to contain an equatorial, thus providing a great saving in cost.

The success of the MMT and other new ideas in telescope construction have led astronomers to think in terms of the "Next Generation Telescope." A Kitt Peak study group is thinking of a 25-m NGT, 5 times the diameter and 25 times the collecting area of the Palomar telescope. They have not chosen between multiple-mirror and other alternative designs. A University of California study group, for their 10-m NGT, is considering especially two designs: a thin single mirror and a honeycomb pattern of mirror segments. A Soviet group is considering a mosaic design in which 500 hexagonal segments make up a 25-m collecting area. A University of Texas group is starting with a 7-m design.

Even as ground-based telescopes increase in number and quality, we have entered the era of the telescope in space. The 0.9-m (36-in) Copernicus space telescope, in which the ultraviolet image was scanned point by point, has now been superseded by the 0.5-m (18-in) International Ultraviolet Explorer (IUE). IUE is operated as a true international observatory, with hundreds of scientists from the United States taking their turns at observing from a control room in Greenbelt, Maryland, and hundreds of European scientists observing from a control room in Villafranca, Spain. Since IUE is in a synchronous orbit, executing a small ellipse over the Atlantic, it is in constant touch with the astronomers. It records ultraviolet spectra on a vidicon screen, allowing much more efficient collection of data than had been possible with Copernicus. The ability to observe faint objects in the ultraviolet for the first time is providing important new data across the board; planets, nebulae, stars, galaxies, quasars have all been observed with IUE.

Another new telescope in space employs an optical principle that is relatively new to astronomers in order to image x-rays. The second High Energy Astronomy Observatory, called the Einstein Observatory, contains four cylindrical telescopes that image x-rays at a grazing angle of less than 1°. Since x-rays would pass through rather than reflect off a mirror at a more normal angle, this grazing technique is necessary to focus x-rays. Cylindrical sections of first a paraboloid and then a hyperboloid focus the x-rays. Only an annulus of optics is seen from the front. Since such a small ring intercepts the incoming radiation, four concentric telescopes are used to increase the collecting area. The telescope, launched in 1979, has returned astounding images of a variety of x-ray sources. By showing that the background of x-rays that had been thought to be general actually originated in quasars too faint to be discovered in other ways, the Einstein Observatory has indicated that the amount of matter in the universe is probably too small to ever end the universe's expansion.

NASA's Space Telescope, with a 2.4-m (95-in) mirror,

should revolutionize many fields of astronomy when it is launched about 1983. It will be above the earth's atmosphere, so will be able to have diffraction-limited images of about 0.1 arc sec even in the ultraviolet, and will have a black sky that will allow the observation of objects about 50 times fainter than the faintest ones that can be observed with ground-based telescopes even at the best sites. As a result, we will be able to expand our observations of classes of objects by a factor of 7 in distance, which is equivalent to a factor of 350 in volume. Applications to improving our understanding of the cosmic distance scale and the structure of the farthest objects in the universe are obvious.

Space Telescope will be launched by Space Shuttle, which will also be able to bring astronauts up to repair and refurbish it from time to time. Space Shuttle will also be used to launch a European Large Infrared Telescope with a 2.8-m mirror. A 1.25-m (50-in) aperture Solar Optical Telescope, which would allow a gain in resolution on the solar surface by a factor of 10, is also on the drawing board for the mid to late 1980's.

This all leads us back to Bell and his book, for the fundamentals about telescopes that he describes are useful for understanding the new telescopes. For example, Hartmann tests, Dawes' limit, and Airy disks are still of importance. Further, Bell recounts the invention of a type of reflecting telescope by the Scottish mathematician James Gregory in 1663, even before Newton suggested his own method of making a reflector. Although Gregorians have not been plentiful in past years, it was realized during the series of Orbiting Solar Observatories that the concentration of sunlight on the secondary mirror that occurs in a Cassegrain can be harmful to the telescope. Thus the telescopes aboard 1980's Solar Maximum Mission and the Solar Optical Telescope are Gregorians. We are back to our roots.

JAY M. PASACHOFF

Williamstown, Massachusetts
November, 1979

REFERENCES

Barlow, B. V., *The Astronomical Telescope* (London: Wykeham Publications Ltd; New York: Springer-Verlag, 1975).

Bowen, I. S., "Future Tools of the Astronomer," *Quarterly Journal of the Royal Astronomical Society,* vol. 8, pp. 9-22, 1967a.

Bowen, I.S., "Astronomical Optics," *Annual Review of Astronomy and Astrophysics,* vol. 5, pp. 45-66, 1967b.

Gascoigne, S. C. B., "Some Recent Advances in the Optics of Large Telescopes," *Quarterly Journal of the Royal Astronomical Society,* vol. 9, pp. 98-115, 1968.

Kuiper, G. P. and Middlehurst, B. M., eds., *Telescopes* (Chicago: University of Chicago Press, 1960). Includes articles on "the 200-Inch Hale Telescope" and "Schmidt Cameras" by I. S. Bowen, "The Lick Observatory 120-Inch Telescope" by W. W. Baustian, and "Design of Reflecting Telescopes" by A. B. Meinel.

Pasachoff, J. M., *Contemporary Astronomy,* 2nd ed. (Philadelphia: Saunders College/HRW, 1981).

Pasachoff, J. M. and Kutner, M. L., *University Astronomy,* (Philadelphia: W. B. Saunders Co., 1978).

Pasachoff, J. M. and Pasachoff, N., "Looking Deeper, Seeing Farther: New Optical Telescopes," *1980 Britannica Yearbook of Science and the Future.*

LARGEST REFLECTING TELESCOPES

TELESCOPE OR INSTITUTION	*LOCATION*	*DIAMETER (m)*	*(in)*	*DATE COMPLETED*	*MIRROR*
Soviet Special Astrophysical Observatory	Caucasus, U.S.S.R.	6.0m	236	1976	Pyrex
Palomar Observatory: Hale Telescope	California	5.0m	200	1950	Pyrex
Multiple Mirror Telescope (MMT)	Arizona	4.5m	176	1979	Fused silica
La Palma Observatory: U.K. Telescope	Canary Islands, Spain	4.2m	166	(1985)	Cer-Vit
Cerro Tololo Inter-American Observatory (KNPO). Mayall Telescope	Chile	4.0m	156	1974	Quartz
Anglo-Australian Telescope (AAT)	Australia	3.9m	153	1975	Cer-Vit
Kitt Peak National Observatory (CTIO)	Arizona	3.8m	150	1975	Cer-Vit
United Kingdom Infrared Telescope (UKIRT)	Hawaii	3.8m	150	1979	Cer-Vit
European Southern Observatory (ESO)	Chile	3.6m	142	1976	Fused quartz

LARGEST REFLECTING TELESCOPES *(contd.)*

TELESCOPE OR INSTITUTION	*LOCATION*	*DIAMETER (m)*	*(in)*	*DATE COMPLETED*	*MIRROR*
Canada-France-Hawaii (CFH)	Hawaii	3.6m	140	1979	Cer-Vit
German-Spanish Astronomical Center, Calar Alto	Spain	3.5m	138	(1983)	Zerodur
NASA-Hawaii	Hawaii	3.0m	120	1979	Cer-Vit
Lick Observatory: Shane Telescope	California	3.0m	120	1959	Pyrex
McDonald Observatory	Texas	2.7m	107	1968	Fused quartz
Crimean Astrophysical Observatory: Shajn Teelscope	Crimea, U.S.S.R.	2.6m	102	1961	Pyrex
Byurakan Observatory	Armenia, U.S.S.R.	2.6m	102	1976	Pyrex
Las Campanas Observatory: Irénée du Pont Telescope	Chile	2.5m	101	1977	Fused quartz
La Palma Observatory: U.K. Isaac Newton Telescope	Canary Islands, Spain	2.5m	101	(1982)	Zerodur
Mt. Wilson Observatory: Hooker Telescope	California	2.5m	100	1917	Glass

TELESCOPE OR INSTITUTION	LOCATION	DIAMETER (m)	(in)	DATE COMPLETED	MIRROR
Space Telescope (ST)	Earth Orbit	2.4m	94	(1983)	(U.L.E. Honeycomb)
Wyoming Infrared Observatory (WIO)	Wyoming	2.3m	92	1977	Cer-Vit
Steward Observatory	Arizona	2.3m	90	1969	Fused quartz
University of Hawaii	Hawaii	2.2m	88	1970	Fused quartz
German-Spanish Astronomical Center, Calar Alto	Spain	2.2m	88	1979	Zerodur

SCHMIDT TELESCOPES

TELESCOPE OR INSTITUTION	LOCATION	DIAMETER CORRECTOR (m)/(in)	DIAMETER MIRROR (m)/(in)	DATE COMPLETED
Karl Schwarzschild Observatory	Germany	1.3m/52in	2.0m/79in	1960
Palomar Observatory	California	1.2m/48in	1.8m/72in	1949
U. K. Schmidt	Australia	1.2m/48in	1.8m/72in	1973
ESO Schmidt	Chile	1.0m/40in	1.6m/63in	1973

PREFACE

This book is written for the many observers, who use telescopes for study or pleasure and desire more information about their construction and properties. Not being a "handbook" in two or more thick quartos, it attempts neither exhaustive technicalities nor popular descriptions of great observatories and their work. It deals primarily with principles and their application to such instruments as are likely to come into the possession, or within reach, of students and others for whom the Heavens have a compelling call.

Much has been written of telescopes, first and last, but it is for the most part scattered through papers in three or four languages, and quite inaccessible to the ordinary reader. For his benefit the references are, so far as is practicable, to English sources, and dimensions are given, regretfully, in English units. Certain branches of the subject are not here discussed for lack of space or because there is recent literature at hand to which reference can be made. Such topics are telescopes notable chiefly for their dimensions, and photographic apparatus on which special treatises are available.

Celestial photography is a branch of astronomy which stands on its own feet, and although many telescopes are successfully used for photography through the help of color screens, the photographic telescope proper and its use belongs to a field somewhat apart, requiring a technique quite its own.

It is many years, however, since any book has dealt with the telescope itself, apart from the often repeated accounts of the marvels it discloses. The present volume contains neither pictures of nebulæ nor speculations as to the habitibility of the planets; it merely attempts to bring the facts regarding the astronomer's chief instrument of research somewhere within grasp and up to the present time.

The author cordially acknowledges his obligations to the important astronomical journals, particularly the Astro-physical Journal, and Popular Astronomy in this country; The Observatory, and the publications of the Royal Astronomical Society in England; the Bulletin de la Société Astronomique de France; and the Astronomische Nachrichten; which, with a few other journals and the official reports of observatories form the body of astronomical knowledge. He also acknowledges the kindness of the various publishers who have extended the courtesy of illustrations, especially Macmillan & Co. and the Clarendon Press, and above all renders thanks to the many friends who have cordially lent a helping hand—the Director and staff of the Harvard Observatory, Dr. George E. Hale, C. A. R. Lundin, manager of the Alvan Clark Corporation, J. B. McDowell, successor of the Brashear Company, H. M. Bennett, the American representative of Carl Zeiss, Jena, and not a few others.

LOUIS BELL.

BOSTON, MASS.,
February, 1922.

CONTENTS

THE EXPLANATION OF ABBREVIATIONS

A. N. = ASTRONOMISCHE NACHRICHTEN.
AP. J. = THE ASTROPHYSICAL JOURNAL.
H. A. = ANNALS OF HARVARD COLLEGE OBSERVATORY.
M. N. = MONTHLY NOTICES OF THE ROYAL ASTRONOMICAL SOCIETY.
OBS. = THE OBSERVATORY.
POP. AST. = POPULAR ASTRONOMY.
ZEIT. F. INST. = ZEITSCHRIFT FÜR INSTRUMENTENKUNDE.

THE TELESCOPE

THE TELESCOPE

CHAPTER I

THE EVOLUTION OF THE TELESCOPE

In the credulous twaddle of an essay on the Lost Arts one may generally find the telescope ascribed to far antiquity. In place of evidence there is vague allusion of classical times or wild flights of fancy like one which argued from the Scriptural statement that Satan took up Christ into a high mountain and showed him all the kingdoms of the earth, that the Devil had a telescope—bad optics and worse theology.

In point of fact there is not any indication that either in classical times, or in the black thousand years of hopeless ignorance that followed the fall of Roman civilization, was there any knowledge of optical instruments worth mentioning.

The peoples that tended their flocks by night in the East alone kept alive the knowledge of astronomy, and very gradually, with the revival of learning, came the spirit of experiment that led to the invention of aids to man's natural powers.

The lineage of the telescope runs unmistakably back to spectacles, and these have an honorable history extending over more than six centuries to the early and fruitful days of the Renaissance.

That their origin was in Italy near the end of the thirteenth century admits of little doubt. A Florentine manuscript letter of 1289 refers to "Those glasses they call spectacles, lately invented, to the great advantage of poor old men when their sight grows weak," and in 1305 Giordano da Rivalto refers to them as dating back about twenty years.

Finally, in the church of Santa Maria Maggiore in Florence lay buried Salvino d'Amarto degli Armati, (obiit 1317) under an epitaph, now disappeared, ascribing to him the invention of spectacles. W. B. Carpenter, F. R. S., states that the inventor tried to keep the valuable secret to himself, but it was discovered and published before his death. At all events the discovery moved swiftly. By the early fourteenth century it had spread to the Low Countries where it was destined to lead to great results, and presently was common knowledge over all civilized Europe.

It was three hundred years, however, between spectacles and the combination of spectacle lenses into a telescope, a lapse of time which to some investigators has seemed altogether mysterious. The ophthalmological facts lead to a simple explanation. The first spectacles were for the relief of presbyopia, the common and lamentable affliction of advancing years, and for this purpose convex lenses of very moderate power sufficed, nor was material variation in power necessary. Glasses having a uniform focus of a foot and a half or thereabouts would serve every practical purpose, but would be no material for telescopes.

Myopia was little known, its acquired form being rare in a period of general illiteracy, and glasses for its correction, especially as regards its higher degrees, probably came slowly and were in very small demand, so that the chance of an optical craftsman having in hand the ordinary convex lenses and those of strong negative curvature was altogether remote. Indeed it was only in 1575 that Maurolycus published a clear description of myopia and hypermetropia with the appropriate treatment by the use of concave and convex lenses. Until both of these, in quite various powers, were available, there was small chance of hitting upon an instrument that required their use in a highly special combination.

At all events there is no definite trace of the discovery of telescopic vision until 1608 and the inventor of record is unquestionably one Jan Lippershey, a spectacle maker of Middelburg in Zeeland, a native of Wesel. On Oct. 2, 1608 the States-General took under consideration a petition which had been presented by Lippershey for a 30-year patent to the exclusive right of manufacture of an instrument for seeing at a distance, or for a suitable pension, under the condition that he should make the instrument only for his country's service.

The States General pricked up its ears and promptly appointed on Oct. 4 a committee to test the new instrument from a tower of Prince Maurice's palace, allotting 900 florins for the purchase of the invention should it prove good. On the 6th the committee reported favorably and the Assembly agreed to give Lippershey 900 florins for his instrument, but desired that it be arranged for use with both eyes.

Lippershey therefore pushed forward to the binocular form and two months later, Dec. 9, he announced his success. On the 15th the new instrument was examined and pronounced good, and the Assembly ordered two more binoculars, of rock crystal, at the same price. They denied a patent on the ground that the invention was known to others, but paid Lippershey liberally as a sort of retainer to secure his exclusive services to the State. In fact even the French Ambassador, wishing to obtain an instrument from him for his King, had to secure the necessary authorization from the States-General.

Bull. de la Soc. Astron. de France.

FIG. 1.—Jan Lippershey, Inventor of the Telescope.

It is here pertinent to enquire what manner of optic tube Lippershey showed to back up his petition, and how it had come to public knowledge. As nearly as we may know these first telescopes were about a foot and a half long, as noted by Huygens, and probably an inch and a half or less in aperture, being constructed of an ordinary convex lens such as was used in spectacles for the aged, and of a concave glass suitable for a bad case of short sightedness, the only kind in that day likely to receive attention.

It probably magnified no more than three or four diameters and was most likely in a substantial tube of firmly rolled, glued, and varnished paper, originally without provision for focussing, since with an eye lens of rather low power the need of adjustment would not be acute.

As to the invention being generally known, the only definite attempt to dispute priority was made by James Metius of Alkmaar, who, learning of Lippershey's petition, on Oct. 17, 1608, filed a similar one, alleging that through study and labor extending over a couple of years he, having accidentally hit upon the idea, had so far carried it out that his instrument made distant objects as distinct as the one lately offered to the States by a citizen and spectacle maker of Middelburg.

He apparently did not submit an instrument, was politely told to perfect his invention before his petition was further considered, and thereafter disappears from the scene, whatever his merits. If he had actually noted telescopic vision he had neither appreciated its enormous importance nor laid the facts before others who might have done so.

The only other contemporary for whom claims have been made is Zacharius Jansen, also a spectacle maker of Middelburg, to whom Pierre Borel, on entirely second hand information, ascribed the discovery of the telescope. But Borel wrote nearly fifty years later, after all the principals were dead, and the evidence he collected from the precarious memories of venerable witnesses is very conflicting and points to about 1610 as the date when Jansen was making telescopes—like many other spectacle makers.[1]

Borel also gave credence to a tale that Metius, seeking Jansen, strayed into Lippershey's shop and by his inquiries gave the shrewd proprietor his first hint of the telescope, but set the date at 1610. A variation of this tale of the mysterious stranger, due to Hieronymus Sirturus, contains the interesting intimation that he may have been of supernatural origin—not further specified. There are also the reports, common among the ignorant or envious, that Lippershey's discovery was accidental, even perhaps made by his children or apprentice.

Just how it actually was made we do not know, but there is no reason to suppose that it was not in the commonplace way

[1] There is a very strong probability that Jansen was the inventor of the compound microscope about the beginning of the seventeenth century.

of experimenting with and testing lenses that he had produced, perhaps those made to meet a vicious case of myopia in one of his patrons.

When the discovery was made is somewhat clearer. Plainly it antedated Oct. 2, and in Lippershey's petition is a definite statement that an instrument had already been tested by some, at least, of the members of the States-General. A somewhat vague and gossipy note in the *Mercure Française* intimates that one was presented to Prince Maurice "about September of the past year" (1608) and that it was shown to the Council of State and to others.

Allowing a reasonable time between Lippershey's discovery and the actual production of an example suitable for exhibition to the authorities, it seems likely that the invention dates back certainly into the summer of 1608, perhaps even earlier.

At all events there is every indication that the news of it spread like wild-fire. Unless Lippershey were unusually careful in keeping his secret, and there are traditions that he was not, the sensational discovery would have been quickly known in the little town and every spectacle maker whose ears it reached would have been busy with it.

If the dates given by Simon Marius in his *Mundus Jovialis* be correct, a Belgian with an air of mystery and a glass of which one of the lenses was cracked, turned up at the Frankfort fair in the autumn of 1608 and at last allowed Fuchs, a nobleman of Bimbach, to look through the instrument. Fuchs noted that it magnified "several" times, but fell out with the Belgian over the price, and returning, took up the matter with Marius, fathomed the construction, tried it with glasses from spectacles, attempted to get a convex lens of longer focus from a Nuremburg maker, who had no suitable tools, and the following summer got a fairly good glass from Belgium where such were already becoming common.

With this Marius eventually picked up three satellites of Jupiter—the fourth awaited the arrival of a superior telescope from Venice. Early in 1609 telescopes "about a foot long" were certainly for sale in Paris, a Frenchman had offered one in Milan by May of that year, a couple of months later one was in use by Harriot in England, an example had reached Cardinal Borghese, and specimens are said to have reached Padua. Fig. 2 from the "*Mundus Jovialis,*" shows Marius with his "Perspicilium," the first published picture of the new instrument. Early in 1610 telescopes were being made in England, but if the few reports of performance, even at this date, are trustworthy, the "Dutch trunk" of that period was of very indifferent quality and power, far from being an astronomical instrument.

One cannot lay aside this preliminary phase of the evolution of the telescope without reference to the alleged descriptions of telescopic apparatus by Roger Bacon, (c. 1270), Giambattista

The Observatory.

FIG. 2.—Simon Marius and his Telescope.

della Porta (1558), and Leonard Digges (1571), details of which may be found in Grant's *History of Physical Astronomy* and many other works.

Of these the first on careful reading conveys strongly the conviction that the author had a pretty clear idea of refraction from the standpoint of visual angle, yet without giving any evidence of practical acquaintance with actual apparatus for doing the things which he suggests.

Given a suitable supply of lenses, it is reasonably certain that Bacon was clever enough to have devised both telescope and microscope, but there is no evidence that he did so, although his manifold activities kept him constantly in public view. It does not seem unlikely, however, that his suggestions in manuscripts, quite available at the time, may have led to the contemporaneous invention of spectacles.

Porta's comments sound like an echo of Bacon's, plus a rather muddled attempt to imagine the corresponding apparatus. Kepler, certainly competent and familiar with the principles of the telescope, found his description entirely unintelligible. Porta, however, was one of the earliest workers on the *camera obscura* and upon this some of his cryptic statements may have borne.

Somewhat similar is the situation respecting Digges. His son makes reference to a manuscript of Roger Bacon as the source of the marvels he describes. The whole account, however, strongly suggests experiments with the *camera obscura* rather than with the telescope.

The most that can be said with reference to any of the three is that, if he by any chance fell upon the combination of lenses that gave telescopic vision, he failed to set down the facts in any form that could be or was of use to others. There is no reason to believe that the Dutch discovery, important as it was, had gone beyond the empirical observation that a common convex spectacle lens and a concave one of relatively large curvature could be placed in a tube, convex ahead, at such a distance apart as to give a clear enlarged image of distant objects.

It remained for Galileo (1564–1647) to grasp the general principles involved and to apply them to a real instrument of research. It was in May 1609 that, on a visit to Venice, he heard reports that a Belgian had devised an instrument which made distant objects seem near, and this being quickly confirmed by a letter from Paris he awakened to the importance of the issue and, returning to Padua, is said to have solved the problem the very night of his arrival.

Next day he procured a plano-convex and a plano-concave lens, fitted them to a lead tube and found that the combination magnified three diameters, an observation which indicates about what it was possible to obtain from the stock of the contemporary spectacle maker.[1] The relation between the power and the foci

[1] The statement by Galileo that he "fashioned" these first lenses can hardly be taken literally if his very speedy construction is to be credited.

of the lenses he evidently quickly fathomed for his next recorded trial reached about eight diameters.

With this instrument he proceeded to Venice and during a month's stay, August, 1609, exhibited it to the senators of the republic and throngs of notables, finally disclosing the secret of its construction and presenting the tube itself to the Doge sitting in full council. This particular telescope was about twenty inches long and one and five eighths inches in aperture, showing plainly that Galileo had by this time found, or more

Courtesy, Macmillan & Co.

FIG. 3.—Galileo.

likely made, an eye lens of short focus, about three inches, quite probably using a well polished convex lens of the ordinary sort as objective

Laden with honors he returned to Padua and settled down to the hard work of development, grinding many lenses with his own hands and finally producing the instrument magnifying some 32 times, with which he began the notable succession of discoveries that laid the foundation of observational astronomy. This with another of similar dimensions is still preserved at the

Galileo Museum in Florence, and is shown in the Frontispiece. The larger instrument is forty-nine inches long and an inch and three quarters aperture, the smaller about thirty-seven inches long and of an inch and five-eighths aperture. The tubes are of paper, the glasses still remain, and these are in fact the first astronomical telescopes.

Galileo made in Padua, and after his return to Florence in the autumn of 1610, many telescopes which found their way over Europe, but quite certainly none of power equalling or exceeding these.

In this connection John Greaves, later Savilian Professor of Astronomy at Oxford, writing from Sienna in 1639, says: "Galileus never made but two good glasses, and those were of old Venice glass." In these best telescopes, however, the great Florentine had clearly accomplished a most workmanlike feat. He had brought the focus of his eye lens down to that usual in modern opera glasses, and has pushed his power about to the limit for simple lenses thus combined.

The lack of clear and homogeneous glass, the great difficulty of forming true tools, want of suitable commercial abrasives, impossibility of buying sheet metals or tubing (except lead), and default of now familiar methods of centering and testing lenses, made the production of respectably good instruments a task the difficulty of which it is hard now to appreciate.

The services of Galileo to the art were of such profound importance, that his form of instrument may well bear his name, even though his eyes were not the first that had looked through it. Such, too, was the judgment of his contemporaries, and it was by the act of his colleagues in the renowned Acaddemia dei Lincei, through the learned Damiscianus, that the name "Telescope" was devised and has been handed down to us.

A serious fault of the Galilean telescope was its very small field of view when of any considerable power. Galileo's largest instrument had a field of but 7′15″, less than one quarter the moon's diameter. The general reason is plain if one follows the rays through the lenses as in Fig. 4 where AB is the distant object, o the objective, e the eye lens, ab the real image in the absence of e, and $a'b'$ the virtual magnified image due to e.

It will be at once seen that the axes of the pencils of rays from all parts of the object, as shown by the heavy lines, act as if they diverged from the optical center of the objective, but diverging still more by refraction through the concave eye lens *e*, fall mostly outside the pupil of the observer's eye. In fact the field depends on the diameter of the pupil of the eye, and on the breadth of the objective.

To the credit of the Galilean form may be set down the convenient erect image, a sharp, if small, field somewhat bettered by a partial compensation of the aberrations of the objective by the concave eye lens, and good illumination. For a distant

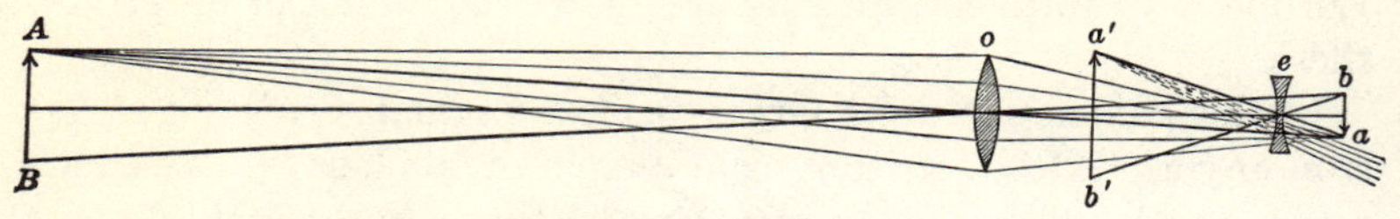

Fig. 4.—Diagram of Galileo's Telescope.

object the lenses were spaced at the difference of their focal lengths, and the magnifying power was the ratio of these, f_o/f_e.

But the difficulty of obtaining high power with a fairly sizeable field was ultimately fatal and the type now survives only in the form of opera and field glasses, usually of 2 to 5 power, and in an occasional negative eye lens for erecting the image in observatory work. Practically all the modern instruments have achromatic objectives and commonly achromatic oculars.

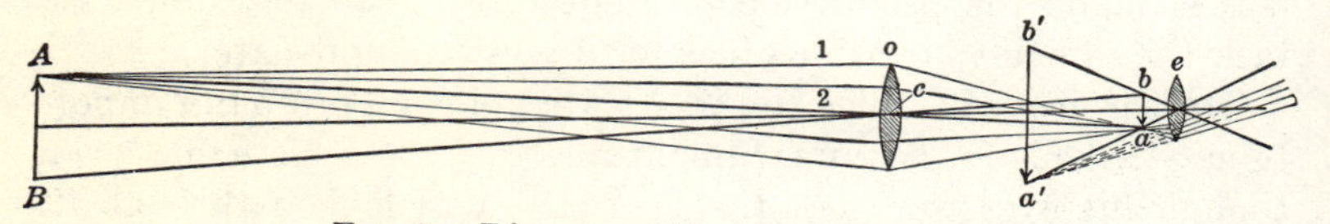

Fig. 5.—Diagram of Kepler's Telescope.

The necessary step forward was made by Johann Kepler (1571–1630), the immortal discoverer of the laws of planetary motion. In his *Dioptrice* (1611) he set forth the astronomical telescope, substantially, save for the changes brought by achromatism, as it has been used ever since. His arrangement was that of Fig. 5 in which the letters have the same significance as in Fig. 4.

There are here three striking differences from the Galilean form. There is a real image in the front focus of the eye lens *e*, the rays passing it are refracted inwards instead of outwards, to the great advantage of the field, and any object placed in the image plane will be magnified together with the image. The first two points Kepler fully realized, the third he probably did not, though it is the basis of the micrometer. The lenses *o* and *e* are obviously spaced at the sum of their focal lengths, and as before the magnifying power is the ratio of these lengths, the visible image being inverted.

Kepler, so far as known, did not actually use the new telescope, that honor falling about half a dozen years later, to Christopher Scheiner, a Jesuit professor of mathematics at Ingolstadt, best known as a very early and most persistent, not to say verbose, observer of sun spots. His *Rosa Ursina* (1630) indicates free use of Kepler's telescope for some years previously, in just what size and power is uncertain.[1] Fontana of Naples also appears to have been early in the field.

But the new instrument despite its much larger field and far greater possibilities of power, brought with it some very serious problems. With increased power came greatly aggravated trouble from spherical aberration and chromatic aberration as well, and the additive aberrations of the eye lens made matters still worse. The earlier Keplerian instruments were probably rather bad if the drawings of Fontana from 1629 to 1636 fairly represent them.

If one may judge from the course of developments, the first great impulse to improvement came with the publication of Descartes' (1596–1650) study of dioptrics in 1637. Therein was set forth much of the theory of spherical aberration and astronomers promptly followed the clues, practical and impractical, thus disclosed.

Without going into the theory of aberrations the fact of importance to the improvement of the early telescope is that the longitudinal spherical aberration of any simple lens is directly proportional to its thickness due to curvature. Hence, other things being equal, the longer the focus for the same aperture the less the spherical aberration both absolutely and relatively to the image. Further, although Descartes knew nothing of chromatic aberration, and the colored fringe about objects seen through the telescope must then have seemed altogether mysterious, it, also, was greatly relieved by lengthening the focus.

[1] Scheiner also devised a crude parallactic mount which he used in his solar observations, probably the first European to grasp the principle of the equatorial. It was only near the end of the century that Roemer followed his example, and both had been anticipated by Chinese instruments with sights.

For the chromatic circle produced by a simple lens of given diameter has a radial width substantially irrespective of the focal length. But increasing the focal length increases in exact proportion the size of the image, correspondingly decreasing the relative effect of the chromatic error.

Descartes also suggested several designs of lenses which would be altogether free of spherical aberration, formed with elliptical or hyperbolic curvature, and for some time fruitless efforts were made to realize this in practice. It was in fact to be near a century before anyone successfully figured non-spherical surfaces. It was spherical quite as much as chromatic aberration that drove astronomers to long telescopes.

Meanwhile the astronomical telescope fell into better hands than those of Scheiner. The first fully to grasp its possibilities was William Gascoigne, a gallant young gentleman of Middleton, Yorkshire, born about 1620 (some say as early as 1612) and who died fighting on the King's side at Marston Moor, July 2, 1644. To him came as early as 1638 the inspiration of utilizing the real focus of the objective for establishing a telescopic sight.

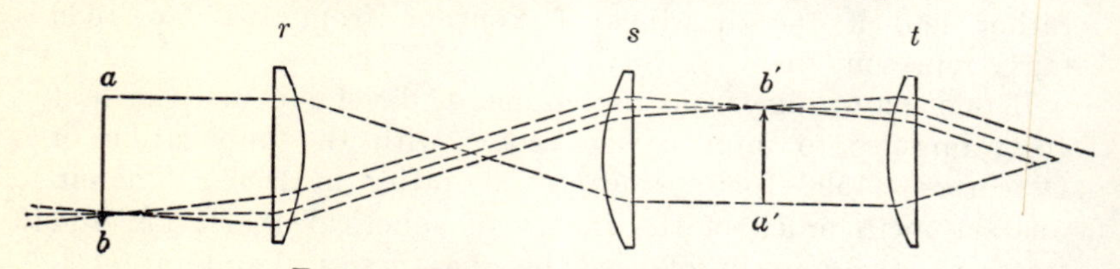

FIG. 6.—Diagram of Terrestrial Ocular.

This shortly took the form of a genuine micrometer consisting of a pair of parallel blades in the focus, moved in opposite directions by a screw of duplex pitch, with a scale for whole revolutions, and a head divided into 100 parts for partial revolutions. With this he observed much from 1638 to 1643, measured the diameters of sun, moon and planets with a good degree of precision, and laid the foundations of modern micrometry. He was equipped by 1639 with what was then called a large telescope.

His untimely death, leaving behind an unpublished treatise on optics, was a grave loss to science, the more since the manuscript could not be found, and, swept away by the storms of war, his brilliant work dropped out of sight for above a score of years.

Meanwhile De Rheita (1597–1660), a Capuchin monk, and an industrious and capable investigator, had been busy with the telescope, and in 1645 published at Antwerp a somewhat bizarre treatise, dedicated to Jesus Christ, and containing not a little practical information. De Rheita had early constructed binoculars, probably quite independently, had lately been diligently experimenting with Descartes' hyperbolic lens, it is needless to say without much success, and was meditating work on a colossal scale—a glass to magnify 4,000 times.

But his real contribution to optics was the terrestrial ocular. This as he made it is shown in Fig. 6 where *a b* is the image

FIG. 7.—Johannes Hevelius.

formed by the objective in front of the eye lens *r*, *s* and *t* two equal lenses separated by their focal lengths and *a′b′* the resultant reinverted image. This form remained in common use until improved by Dollond more than a century later.

A somewhat earlier form ascribed to Father Scheiner had merged the two lenses forming the inverting system of Fig. 6, into a single lens used at its conjugate foci.

Closely following De Rheita came Johannes Hevelius (1611–1687) of Danzig, one of the really important observers of the seventeenth century. His great treatise *Selenographia* published in 1647 gives us the first systematic study of the moon, and a brief but illuminating account of the instruments of the time and their practical construction.

At this time the Galilean and Keplerian forms of telescope were in concurrent use and Hevelius gives directions for designing and making both of them. Apparently the current instruments were not generally above five or six feet long and from Hevelius' data would give not above 30 diameters in the Galilean form. There is mention, however, of tubes up to 12 feet in length, and of the advantage in clearness and power of the longer focus plano-convex lens. Paper tubes, evidently common, are condemned, also those of sheet iron on account of their weight, and wood was to be preferred for the longer tubes.

Evidently Hevelius had at this time no notion of the effect of the plano-convex form of lens as such in lessening aberration, but he mentions a curious form of telescope, actually due to De Rheita, in which the objective is double, apparently of two plano-convex lenses, the weaker ahead, and used with a concave eye lens. If properly proportioned such a doublet would have less than a quarter the spherical aberration of the equivalent double convex lens.

Hevelius also mentions the earlier form of reinverting telescope above referred to, and speaks rather highly of its performance. To judge from his numerous drawings of the moon made in 1643 and 1644, his telescopes were much better than those of Scheiner and Fontana, but still woefully lacking in sharp definition.

Nevertheless the copper plates of the *Selenographia*, representing every phase of the moon, placed the lunar details with remarkable accuracy and formed for more than a century the best lunar atlas available. One acquires an abiding respect for the patience and skill of these old astronomers in seeing how much they did with means utterly inadequate.

One may get a fair idea of the size, appearance, and mounting of telescopes in this early day from Fig. 8, which shows a somewhat advanced construction credited by Hevelius to a suggestion in Descartes' *Dioptrica*. Appearances indicate that the tube was somewhere about six feet long, approximately two inches in aperture, and that it had a draw tube for focussing. The offset head of the mount to allow observing near the zenith is worth an extra glance.

Incidentally Hevelius, with perhaps pardonable pride, also explains the "Polemoscope," a little invention of his own, made, he tells us, in 1637. It is nothing else than the first periscope, constructed as shown in Fig. 9, a tube *c* with two right angled

Fig. 8.—A Seventeenth Century Astronomer and his Telescope.

branches, a fairly long one *e* for the objective *f*, a 45° mirror at *g*, another at *a*, and finally the concave ocular at *b*. It was of modest size, of tubes 1⅔ inch in diameter, the longer tube being 22 inches and the upper branch 8 inches, a size well suited for trench or parapet.

Even in these days of his youth Hevelius had learned much of practical optics as then known, had devised and was using very rational methods of observing sun-spots by projection in a darkened room, and gives perhaps the first useful hints at testing telescopes by such solar observations and on the planets. He was later to do much in the development and mounting of long telescopes and in observation, although, while progressive in other respects, he very curiously never seemed to grasp the importance of telescopic sights and consistently refused to use them.

Telescope construction was now to fall into more skillful hands. Shortly after 1650 Christian Huygens (1629–1695), and his accomplished brother Constantine awakened to a keen interest in astronomy and devised new and excellent methods of forming accurate tools and of grinding and polishing lenses.

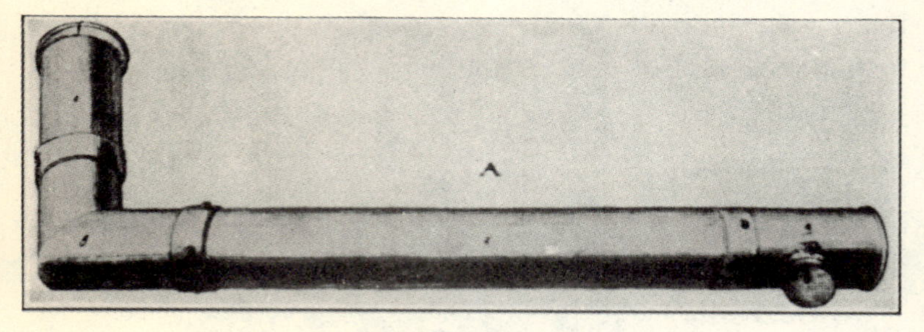

FIG. 9.—The first Periscope.

By 1655 they had completed an instrument of 12 feet focus with which the study of Saturn was begun, Titan the chief satellite discovered, and the ring recognized. Pushing further, they constructed a telescope of 23 feet focal length and 2⅓ inches aperture, with which four years later Christian Huygens finally solved the mystery of Saturn's ring.

Evidently this glass, which bore a power of 100, was of good defining quality, as attested by a sketch of Mars late in 1695 showing plainly Syrtis Major, from observation of which Huygens determined the rotation period to be about 24 hours.

The Huygens brothers were seemingly the first fully to grasp the advantage of very long focus in cutting down the aberrations, the aperture being kept moderate. Their usual proportions were about as indicated above, the aperture being kept somewhere nearly as the square root of the focus in case of the larger glasses.

In the next two decades the focal length of telescopes was pushed by all hands to desperate extremes. The Huygens brothers extended themselves to glasses up to 210 feet focus and built many shorter ones, a famous example of which, of 6 inches aperture and 123 feet focal length, presented to the Royal Society, is still in its possession. Auzout produced even longer telescopes, and Divini and Campani, in Rome, of whom the last named made Cassini's telescopes for the Observatory of Paris, were not far behind. The English makers were similarly busy, and Hevelius in Danzig was keeping up the record.

Fig. 10.—Christian Huygens.

Clearly these enormously long telescopes could not well be mounted in tubes and the users were driven to aerial mountings, in which the objective was at the upper end of a spar or girder and the eye piece at the lower. Figure 11 shows an actual construction by Hevelius for an objective of 150 feet focal length.

In this case the main support was a T beam of wooden planks well braced together. Additional stiffness was given by light wooden diaphragms at short intervals with apertures of about 8 inches next to the objective, and gradually increasing downwards. The whole was lined up by equalizing tackle in the vertical plane, and spreaders with other tackle at the joints of the 40-

foot sections of the main beam. The mast which supported the whole was nearly 90 feet high.

So unwieldly and inconvenient were these long affairs that, quite apart from their usual optical imperfections, it is little wonder that they led to no results commensurate with their size. In fact nearly all the productive work was done with telescopes from 20 to 35 feet long, with apertures roughly between 2 and 3 inches.

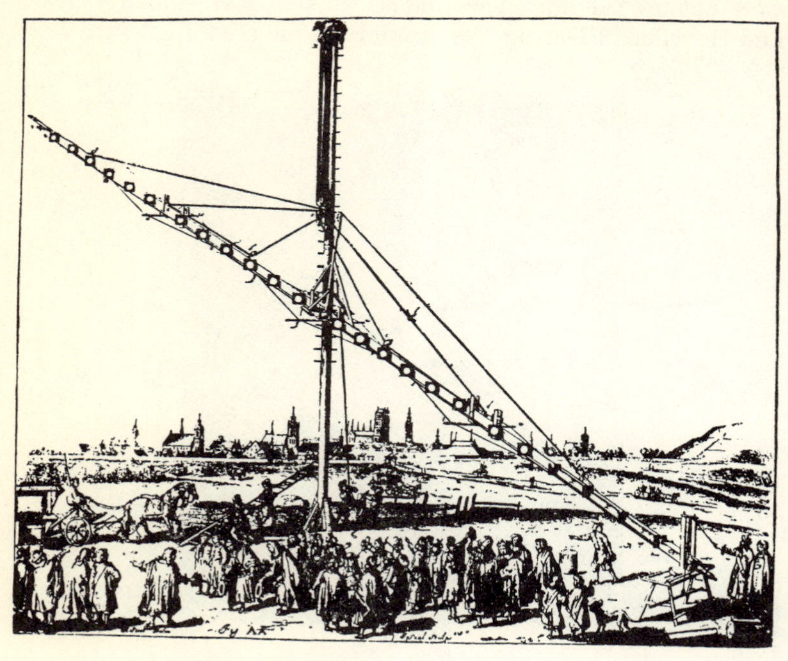

Fig. 11.—Hevelius' 150-foot Telescope.

Dominique Cassini to be sure, scrutinizing Saturn in 1684 with objectives by Campani, of 100 and 136 feet focus picked up the satellites Tethys and Dione, but he had previously found Iapetus with a 17-foot glass, and Rhea with one of 34 feet. The longer glasses above mentioned had aerial mounts but the smaller ones were in tubes supported on a sort of ladder tripod. A 20-foot telescope, power 90, gave Cassini the division in Saturn's ring.

A struggle was still being kept up for the non-spherical curves urged by Descartes. It is quite evident that Huygens had a go at them, and Hevelius thought at one time that he had mastered the hyperbolic figure, but his published drawings give no indication that he had reduced spherical aberration to any perceptible degree. At this time the main thing was to get good glass and give it true figure and polish, in which Huygens and Campani excelled, as the work on Saturn witnesses.

These were the days of the dawn of popular astronomy and many a gentleman was aroused to at least a casual interest in observing the Heavens. Notes Pepys in his immortal *Diary:* "I find Reeves there, it being a mighty fine bright night, and so upon my leads, though very sleepy, till one in the morning, looking on the moon and Jupiter, with this twelve foot glass, and another of six foot, that he hath brought with him to-night, and the sights mighty pleasant, and one of the glasses I will buy."

Little poor Pepys probably saw, by reason of his severe astigmatism, but astronomy was in the air with the impulse that comes to every science after a period of brilliant discovery. Another such stimulus came near the end of the eighteenth century, with the labors of Sir William Herschel.

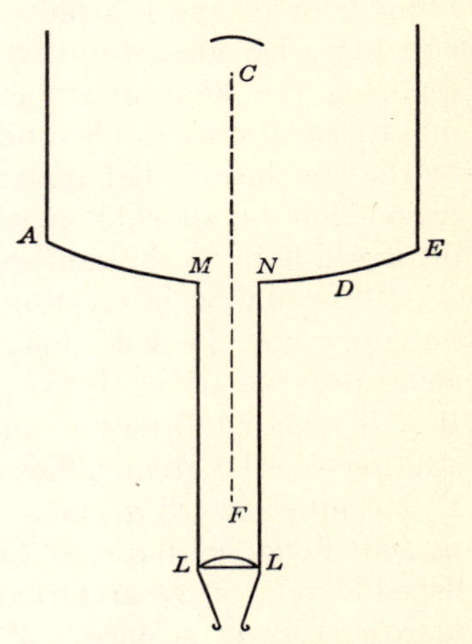

Fig. 12.—Gregory's Diagram of his Telescope.

Just at this juncture comes one of the interesting episodes of telescopic history, the ineffectual and abandoned experiments on reflecting instruments.

In 1663 James Gregory (1638–1675) a famous Scottish mathematician, published his *Optica Promota*, in which he described the rather elegant construction which bears his name, a perforated parabolic mirror with an elliptical mirror forward of the focus returning an image to the ocular through the perforation. It was convenient in that it gave an erect image, and it was sound theoretically, and, as the future proved, practically, but the curves were quite too much for the contemporary opticians. Figure 12 shows the diagrammatic construction as published.

The next year Gregory started Reive, a London optician,

doubtless the same mentioned by Pepys, on the construction of a 6-foot telescope. This rather ambitious effort failed of material success through the inability of Reive to give the needed figures to the mirrors,[1] and of it nothing further appears until the ingenious Robert Hooke (1635–1703) executed in 1674 a Gregorian, apparently without any notable results. There is a well defined tradition that Gregory himself was using one in 1675, at the time of his death, but the invention then dropped out of sight.

No greater influence on the art attended the next attempt at a reflector, by Isaac Newton (1643–1727). This was an early outcome of his notable discovery of the dispersion of light by prisms, which led him to despair of improving refracting telescopes and turned his mind to reflectors.

Unhappily in an experiment to determine whether refraction and dispersion were proportional he committed the singular blunder of raising the refractive index of a water-filled prism to equality with glass by dissolving sugar of lead in it. Without realizing the impropriety of thus varying two quite unknown quantities at once in his crucial experiment, he promptly jumped to the conclusion that refraction and dispersion varied in exact proportion in all substances, so that if two prisms or lenses dispersed light to the same extent they must also equally refract it. It would be interesting to know just how the fact of his bungling was passed along to posterity. As a naïve apologist once remarked, it was not to be found in his "*Optics*." But Sir David Brewster and Sir John Herschel, both staunch admirers of the great philosopher, state the fact very positively. If one may hazard a guess it crept out at Cambridge and was passed along, perhaps to Sir William Herschel, via the unpublished history of research that is rich in picturesque details of the mare's nests of science. At all events a mistake with a great name behind it carries far, and the result was to delay the production of the achromatic telescope by some three quarters of a century.

Turning from refractors he presented to the Royal Society just after his election as Fellow in 1672, the little six-inch model of his device which was received with acclamation and then lay on the shelf without making the slightest impression on the art, for full half a century.

[1] He attempted to polish them on cloth, which in itself was sufficient to guarantee failure.

Newton, by dropping the notion of direct view through the tube, hit upon by far the simplest way of getting the image outside it, by a plane mirror a little inside focus and inclined at 45°, but injudiciously abandoned the parabolic mirror of his original paper on dispersion. His invention therefore as actually made public was of the combination with a spherical concave mirror of a plane mirror of elliptical form at 45°, a construction which in later papers he defended as fully adequate.[1]

FIG. 13.—Newton's Model of his Reflector.

His error in judgment doubtless came from lack of practical astronomical experience, for he assumed that the whole real trouble with existing telescopes was chromatic aberration, which in fact worried the observer little more than the faults due to other causes, since the very low luminosity toward the ends

[1] In Fig 13, *A* is the support of the tube and focussing screw, *B* the main mirror, an inch in diameter, *CD* the oblique mirror, *E* the principal focus, *F* the eye lens, and *G* the member from which the oblique mirror is carried.

of the spectrum enormously lessens the indistinctness due to dispersion.

As a matter of fact the long focus objective of small aperture did very creditable work, and its errors would not compare unfavorably with those of a spherical concave mirror of the wide aperture planned by Newton. Had he actually made one of his telescopes of fair dimensions and power the definition would infallibly have been wrecked by the aberrations due to spherical figure.[1]

It is quite likely that appreciation of this, and the grave doubts of both Newton and Huygens as to obtaining a proper parabolic curve checked further developments. About the beginning of

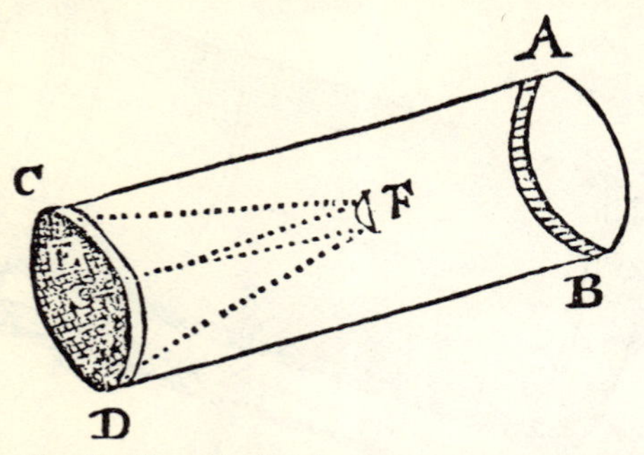

FIG. 14.—De Bercé's sketch of Cassegrain's Telescope.

the year 1672 M. Cassegrain communicated to M. de Bercé a design for a reflecting telescope, which eventually found its way into the *Philosophical Transactions* of May in that year, after previous publication in the *Journal des Sçavans*. Figure 14 shows de Bercé's rough original sketch. It differed from Gregory's construction in that the latter's elliptical concave mirror placed outside the main focus, was replaced by a convex mirror placed inside focus. The image was therefore inverted.

The inventor is referred to in histories of science as "Cassegrain, a Frenchman." He was in fact Sieur Guillaume Cassegrain, sculptor in the service of Louis Quatorze, modeller and founder of many statues. In 1666 he was paid 1200 livres for executing

[1] In fact a "four foot telescope of Mr. Newton's invention" brought before the Royal Society two weeks after his original paper, proved only fair in quality, was returned somewhat improved at the next meeting, and then was referred to Mr. Hooke to be perfected as far as might be, after which nothing more was heard of it.

a bust of the King modelled by Bertin, and later made many replicas from the antique for the decoration of His Majesty's gardens at Versailles. He disappeared from the royal records in 1684 and probably died within a year or two of that date.

At the period here concerned he apparently, like de Bercé, was of Chartres. Familiar with working bronzes and with the art of the founder, he was a very likely person to have executed specula. Although there is no certainty that he actually made a telescope, a contemporary reference in the *Journal des Sçavans* speaks of his invention as a "petite lunette d'approche," and one does not usually suggest the dimensions of a thing non-existent. How long he had been working upon it prior to the period about the beginning of 1672 when he disclosed the device to de Bercé is unknown.

Probably Newton's invention was the earlier, but the two were independent, and it was somewhat ungenerous of Newton to criticise Cassegrain, as he did, for using spherical mirrors, on the strength of de Bercé's very superficial description, when he himself considered the parabolic needless.

However, nothing further was done, and the devices of Gregory, Newton and Cassegrain went together into the discard for some fifty years.

These early experiments gave singularly little information about material for mirrors and methods of working it, so little that those who followed, even up to Lord Rosse, had to work the problems out for themselves. We know from his original paper that Newton used bell-metal, whitened by the addition of arsenic, following the lore of the alchemists.

These speculative worthies used to alloy copper with arsenic, thinking that by giving it a whitish cast they had reached a sort of half way point on the road to silver. Very silly at first thought, but before the days of chemical analysis, when the essential properties of the metals were unknown, the way of the scientific experimenter was hard.

What the "steely matter, imployed in London" of which Newton speaks in an early paper was, we do not know—very likely one of the hard alloys much richer in tin than is ordinary bell-metal. Nor do we know to what variety of speculum metal Huygens refers in his correspondence with Newton.

As to methods of working it Newton only disclosed his scheme of pitch-polishing some thirty years after this period, while it is a matter of previous record, that Huygens had been in the habit of polishing his true tools on pitch from some date unknown. Probably neither of them originated the practice. Opticians are a peculiarly secretive folk and shop methods are likely to be kept for a long time before they leak out or are rediscovered.

Modern speculum metal is substantially a definite compound of four atoms copper and one tin ($SnCu_4$), practically 68 per cent copper and 32 per cent tin, and is now, as it was in all previous modifications, a peculiarly mean material to cast and work. Thus exit the reflector.

The long telescope continued to grow longer with only slow improvement in quality, but the next decade was marked by the introduction of Huygens' eyepiece, an immense improvement over the single lens which had gone before, and with slight modifications in use today.

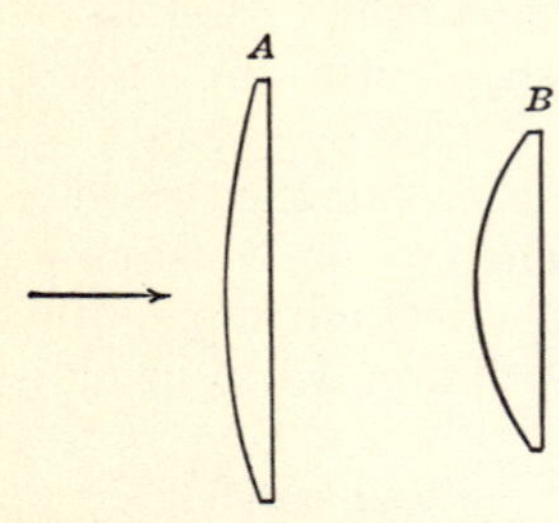

Fig. 15.—Diagram of Huygens' Eyepiece.

This is shown in section in Fig. 15. It consists of a field lens *A*, plano-convex, and an eye lens *B* of one-third the focal length, the two being placed at the difference of their focal lengths apart with (in later days) a stop half way between them. The eye piece is pushed inside the main focus until the rays which fall on the field lens focus through the eye lens.

The great gain from Huygens' view-point was a very much enlarged clear field—about a four-fold increase—and in fact the combination is substantially achromatic, particularly important now when high power oculars are needed.

Still larger progress was made in giving the objective a better form with respect to spherical aberration, the "crossed" lens being rather generally adopted. This form is double convex, and if of ordinary glass, with the rear radius six times the front radius, and gives even better results than a plano-convex in its best position-plane side to the rear. Objectives were rated on focal length for the green rays, that is, the bright central part of the spectrum, the violet rays of course falling short and the red running beyond.

To give customary dimensions, a telescope of 3 inches aperture, with magnifying power of 100, would be of about 30 feet focus with the violet nearly 6 inches short and the red a similar amount long. It is vast credit to the early observers that with such slender means they did so much. But in fact the long telescope had reached a mechanical *impasse*, so that the last quarter of the seventeenth century and the first quarter of the next were marked chiefly by the development of astronomy of position with instruments of modest dimensions.

In due time the new order came and with astounding suddenness. Just at the end of 1722 James Bradley (1692–1762)

Fig. 16.—The First Reflector. John Hadley, 1722.

measured the diameter of Venus with an objective of 212 ft. 3 in. focal length; about three months later John Hadley (1682–1744) presented to the Royal Society the first reflecting telescope worthy the name, and the old order practically ended.

John Hadley should in fact be regarded as the real inventor of the reflector in quite the same sense that Mr. Edison has been held, *de jure* and *de facto*, the inventor of the incandescent electric lamp. Actually Hadley's case is the stronger of the two, for the only things which could have been cited against him were abandoned experiments fifty years old. Moreover he took successfully the essential step at which Gregory and Newton had stumbled or turned back—parabolizing his speculum.

The instrument he presented was of approximately 6 inches aperture and 62⅝ inches focal length, which he had made and tested some three years previously; on a substantial altazimuth mount with slow motions. He used the Newtonian oblique mirror and the instrument was provided with both convex and concave eye lenses, with magnifications up to about 230.

The whole arrangement is shown in Fig. 16 which is for the most part self explanatory. It is worth noting that the speculum is positioned in the wooden tube by pressing it forward against three equidistant studs by three corresponding screws at the rear, that a slider moved by a traversing screw in a wide groove carries the small mirror and the ocular, that there is a convenient door for access to the mirror, and also a suitable finder. The motion in altitude is obtained by a key winding its cord against gravity. That in azimuth is by a roller support along a horizontal runway carried by an upright, and is obtained by the key with a cord pull off in one direction, and in the other, by springs within the main upright, turning a post of which the head carries cheek pieces on which rest the trunnions of the tube.

A few months later this telescope was carefully tested, by Bradley and the Rev. J. Pound, against the Huygens objective of 123 feet focus possessed by the Royal Society, and with altogether satisfactory results. Hadley's reflector would show everything which could be seen by the long instrument, bearing as much power and with equal definition, though somewhat lessened light. In particular they saw all five satellites of Saturn, Cassini's division, which the inventor himself had seen the previous year even in the northern edge of the ring beyond the planet, and the shadow of the ring upon the ball.

The casting of the large speculum was far from perfect, with many spots that failed to take polish, but the figure must have been rather good. A spherical mirror of these dimensions would give an aberration blur something like twenty times the width of Cassini's division, and the chance of seeing all five satellites with it would be negligibly small.

Further, Hadley presently disclosed to others not only the method he used in polishing and parabolizing specula, but his method of testing for true figure by the aberrations disclosed as he worked the figure away from the sphere—a scheme frequently used even to this day.

The effect of Hadley's work was profound. Under his guidance others began to produce well figured mirrors, in particular Molyneux and Hawksbee; reflecting telescopes became fairly common; and in the beginning of the next decade James Short, (1710-1768), possessed of craftsmanship that approached wizardry, not only fully mastered the art of figuring the paraboloid, but at once took up the Gregorian construction with its ellipsoidal small mirror, with much success.

His specula were of great relative aperture, F/4 to F/6, and from the excellent quality of his metal some of them have retained their fine polish and definition after more than a century. He is said to have gone even up to 12 inches in diameter. His exact methods of working died with him. Even his tools he ordered to be destroyed before his death.

The Cassegrain reflector, properly having a parabolic large mirror and a hyperbolic small one, seems very rarely to have been made in the eighteenth century, though one certainly came into the hands of Ramsden (1735–1800).

Few refractors for astronomical use were made after the advent of the reflector, which was, and is, however, badly suited for the purposes of a portable spy-glass, owing to trouble from stray light. The refractor therefore permanently held its own in this function, despite its length and uncorrected aberrations.

Relief was near at hand, for hardly had Short started on his notable career when Chester Moor Hall, Esq. (1704–1771) a gentleman of Essex, designed and caused to be constructed the first achromatic telescope, with an objective of crown and flint glass. He is stated to have been studying the problem for several years, led to it by the erroneous belief (shared by Gregory long before) that the human eye was an example of an achromatic instrument.

Be this as it may, Hall had his telescopes made by George Bast of London at least as early as 1733, and according to the best available evidence several instruments were produced, one of them of above 2 inches aperture on a focal length of about 20 inches (F/8) and further, subsequently such instruments were made and sold by Bast and other opticians.

These facts are clear and yet, with knowledge of them among London workmen as well as among Hall's friends, the invention made no impression, until it was again brought to light, and patented, by the celebrated John Dollond (1706–1761) in the year 1758.

Physical considerations give a clue to this singular neglect. The only glasses differing materially in dispersion available in

Courtesy, Macmillan & Co.

Fig. 17.—John Dollond.

Hall's day were the ordinary crown, and such flint as was in use in the glass cutting trade,—what we would now know as a light flint, and far from homogeneous at that.

Out of such material it was practically very hard (as the Dollonds quickly found) to make a double objective decently free from spherical aberration, especially for one working, as Hall quite assuredly did, by rule of thumb. With the additional handicap of flint full of faults it is altogether likely that these first achromatics, while embodying the correct principles, were not good enough to make effective headway against the cheaper and simpler spy-glass of the time.

Dollond, although in 1753 he strongly supported Newton's error in a Royal Society paper against Euler's belief in achromatism, shifted his view a couple of years later and after a considerable period of skilful and well ordered experimenting published his discovery of achromatism early in 1758, for which a patent was granted him April 19, while in the same year the Royal Society honored him with the Copley medal. From that time until his death, late in 1761, he and his son Peter Dollond (1730–1820) were actively producing achromatic glasses.

The Dollonds were admirable craftsmen and their early product was probably considerably better than were Hall's objectives but they felt the lack of suitable flint and soon after John Dollond's death, about 1765, the son sought relief in the triple objective of which an early example is shown in Fig. 18, and which, with some modifications, was his standard form for many years.

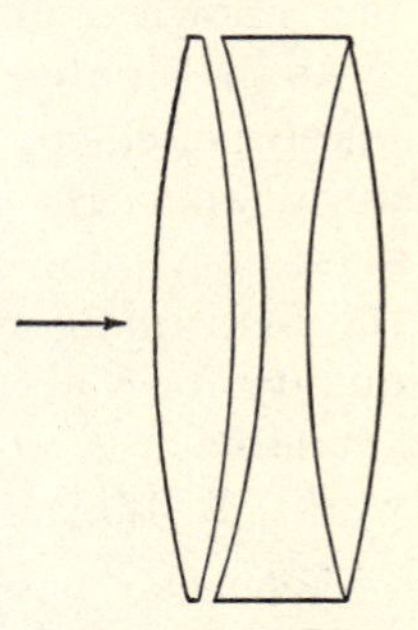

FIG. 18.—Peter Dollond's Triple Objective.

Other opticians began to make achromatics, and, Peter Dollond having threatened action for infringement, a petition was brought by 35 opticians of London in 1764 for the annulment of John Dollond's patent, alleging that he was not the original inventor but had knowledge of Chester Moor Hall's prior work. In the list was George Bast, who in fact did make Hall's objectives twenty five years before Dollond, and also one Robert Rew of Cold-bath Fields, who claimed in 1755 to have informed Dollond of the construction of Hall's objective.

This was just the time when Dollond came to the right about face on achromatism, and it may well be that from Rew or elsewhere he may have learned that a duplex achromatic lens had really been produced. But his Royal Society paper shows that his result came from honest investigations, and at worst he is in about the position of Galileo a century and a half before.

The petition apparently brought no action, perhaps because Peter Dollond next year sued Champneys, one of the signers, and obtained judgment. It was in this case that the judge (Lord Camden) delivered the oft quoted dictum: "It was not the person who locked up his invention in his scrutoire that ought to profit by a patent for such invention, but he who brought it forth for the benefit of the public.[1]"

This was sound equity enough, assuming the facts to be as stated, but while Hall did not publish the invention admittedly made by him, it had certainly become known to many. Chester Moor Hall was a substantial and respected lawyer, a bencher of the Inner Temple, and one is inclined to think that his alleged concealment was purely constructive, in his failing to contest Dollond's claim.

Had he appeared at the trial with his fighting blood up, there is every reason to believe that he could have established a perfectly good case of public use quite aside from his proof of technical priority. However, having clearly lost his own claims through *laches*, he not improbably was quite content to let the tradesmen fight it out among themselves. Hall's telescopes were in fact known to be in existence as late as 1827.

As the eighteenth century drew toward its ending the reflecting telescope, chiefly in the Gregorian form, held the field in astronomical work, the old refractor of many draw tubes was the spyglass of popular use, and the newly introduced achromatic was the instrument of "the exclusive trade." No glass of suitable quality for well corrected objectives had been produced, and that available was not to be had in discs large enough for serious work. A 3-inch objective was reckoned rather large.

[1] Commonly, but it appears erroneously, ascribed to Lord Mansfield.

CHAPTER II

THE MODERN TELESCOPE

The chief link between the old and the new, in instrumental as well as observational astronomy, was Sir William Herschel (1738–1822). In the first place he carried the figuring of his mirrors to a point not approached by his predecessors, and second, he taught by example the immense value of aperture in definition and grasp of light. His life has never been adequately written, but Miss Clerke's "*The Herschels and Modern Astronomy*" is extremely well worth the reading as a record of achievement that knew not the impossible.

Courtesy, Macmillan & Co.

FIG. 19.—Sir William Herschel.

He was the son of a capable band-master of Hanover, brought up as a musician, in a family of exceptional musical abilities, and in 1757 jumped his military responsibilities and emigrated to England, to the world's great gain. For nearly a decade he struggled upward in his art, taking meanwhile every opportunity for self-education, not only in the theory of music but in mathematics and the languages, and in 1767 we find him settled in fashionable Bath, oboist in a famous orchestra, and organist of the Octagon Chapel. His abilities brought him many pupils, and ultimately he became director of the orchestra in which he had played, and the musical dictator of the famous old resort.

In 1772 came his inspiration in the loan of a 2-foot Gregorian reflector, and a little casual star-gazing with it. It was the opening of the kingdom of the skies, and he sought to purchase a telescope of his own in London, only to find the price too great for his means. (Even a 2-foot, of 4½ inches aperture, by Short was listed at five-and-thirty guineas.) Then after some futile attempts at making a plain refractor he settled down to hard work at casting and polishing specula.

Although possessed of great mechanical abilities the difficult technique of the new art long baffled him, and he cast and worked some 200 small discs in the production of his first successful telescopes, to say nothing of a still greater number in larger sizes in his immediately subsequent career.

As time went on he scored a larger proportion of successes, but at the start good figure seems to have been largely fortuitous. Inside of a couple of years, however, he had mastered something of the art and turned out a 5-foot instrument which seems to have been of excellent quality, followed later by a 7-foot (aperture 6¼ inches) even better, and then by others still bigger.

The best of Herschel's specula must have been of exquisite figure. His 7-foot was tested at Greenwich against one of Short's of 9½ inches aperture much to the latter's disadvantage. His discovery with the 7-foot, of the "Georgium Sidus" (Uranus) in 1781 won him immediate fame and recognition, beside spurring him to greater efforts, especially in the direction of larger apertures, of which he had fully grasped the importance.

In 1782 he successfully completed a 12-inch speculum of 20 feet focus, followed in 1788 by an 18-inch of the same length. The previous year he first arranged his reflector as a "front view" telescope—the so-called Herschelian. Up to this time he, except for a few Gregorians, had used Newton's oblique mirror.

The heavy loss of light (around 40 per cent) in the second reflection moved him to tilt the main mirror so as to throw the focal point to the edge of the aperture where one could look downward upon the image through the ocular as shown in Fig. 20. Here *SS* is the great speculum, *O* the ocular and *i* the image formed near the rim of the tube. In itself the tilting would seriously impair the definition, but Herschel wisely built his telescopes of moderate relative aperture (F/10 to F/20), so that this difficulty was considerably lessened, while the saving of light, amounting to nearly a stellar magnitude, was important.

Meanwhile he was hard at work on his greatest mirror, of 48 inches clear aperture and 40 feet focal length, the father of the great line of modern telescopes. It was finished in the summer of 1789. The speculum was 49½ inches in over-all diameter, 3½ inches thick and weighed as cast 2118 lbs. The completion of this instrument, which would rank as large even today, was made notable by the immediate discovery of two new satellites of Saturn, Enceladus and Mimas.

It also proved of very great value in sweeping for nebulæ, but its usefulness seems to have been much limited by the flexure of the mirror under its great weight, and by its rapid tarnishing. It required repolishing, which meant refiguring, at least every two years, a prodigious task.[1]

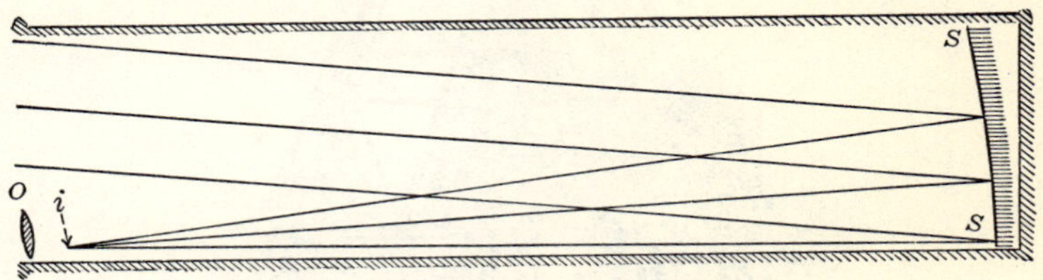

FIG. 20.—Herschel's Front View Telescope.

It was used as a front view instrument and was arranged as shown in Fig. 21. Obviously the front view form has against it the mechanical difficulty of supporting the observer up to quite the full focal length of the instrument in air, a difficulty vastly increased were the mount an equatorial one, so that for the great modern reflectors the Cassegrain form, looked into axially upward, and in length only a third or a quarter of the equivalent focus, is almost universal.

As soon as the excellent results obtained by Herschel became generally known, a large demand arose for his telescopes, which he filled in so far as he could spare the time from his regular

[1] This was probably due not only to unfavorable climate, but to the fact that Herschel, with all his ingenuity, does not appear to have mastered the casting difficulty, and was constrained to make his big speculum of Cu 75 per cent, Sn 25 per cent, a composition working rather easily and taking beautiful, but far from permanent, polish. He never seems to have used practically the $SnCu_4$ formula, devised empirically by Mudge (Phil. Trans. **67,** 298), and in quite general use thereafter up to the present time.

work, and not the least of his services to science was the distribution of telescopes of high quality and consequent strong stimulus to general interest in astronomy.

Two of his instruments, of 4- and 7-feet focus respectively, fell into the worthy hands of Schröter at Lilienthal and did sterling service in making his great systematic study of the lunar surface. At the start even Herschel's 7-foot telescope brought 200 guineas, and the funds thus won he promptly turned to research.

Courtesy, Macmillan & Co.

Fig. 21.—Herschel's Forty-foot Telescope.

We sometimes think of the late eighteenth century as a time of license unbounded and the higher life contemned, but Herschel wakened a general interest in unapplied science that has hardly since been equalled and never surpassed. Try to picture social and official Washington rushing to do honor to some astronomer who by luck had found the trans-Neptunian planet; the diplomatic corps crowding his doors, and his very way to the Naval Observatory blocked by the limousines of the curious and admiring, and some idea may be gained of what really happened to the unassuming music master from Bath who suddenly found himself famous.

Great as were the advances made by Herschel the reflector

was destined to fall into disuse for many years. The fact was that the specula had to be refigured, as in the case of the great 40-foot telescope, quite too often to meet the requirements of the ordinary user, professional or amateur. Only those capable of doing their own figuring could keep their instruments conveniently in service.

Sir W. Herschel always had relays of specula at hand for his smaller instruments, and when his distinguished son, Sir John F. W. Herschel, went on his famous observing expedition to the Cape of Good Hope in 1834–38 he took along his polishing machine and three specula for his 20-foot telescope. And he needed them indeed, for a surface would sometimes go bad even in a week, and regularly became quite useless in 2 or 3 months.

Makers who used the harder speculum metal, very brittle and scarcely to be touched by a file, fared better, and some small mirrors, well cared for, have held serviceable polish for many years. Many of these instruments of Herschel's time, too, were of very admirable performance.

Some of Herschel's own 7-foot telescopes give evidence of exquisite figure and he not only commonly used magnifying powers up to some 80 per inch of aperture, a good stiff figure for a telescope old or new, but went above 2,000, even nearly to 6,000 on one of his 6½-inch mirrors without losing the roundness of the star image. "Empty magnification" of course, gaining no detail whatever, but evidence of good workmanship.

Many years later the Rev. W. R. Dawes, the famous English observer, had a 5-inch Gregorian, commonly referred to as "The Jewel," on which he used 430 diameters, and pushed to 2,000 on Polaris without distortion of the disc. Comparing it with a 5-foot (approximately 4-inch aperture) refractor, he reports the Gregorian somewhat inferior in illuminating power; "But in sharpness of definition, smallness of discs of stars, and hardness of outline of planets it is superior." All of which shows that while methods and material may have improved, the elders did not in the least lack skill.

The next step forward, and a momentous one, was to be taken in the achromatic refractor. Its general principles were understood, but clear and homogeneous glass, particularly flint glass, was not to be had in pieces of any size. "Optical glass," as we understand the term, was unknown.

It is a curious and dramatic fact that to a single man was due not only the origin of the art but the optical glass industry of the world. If the capacity for taking infinite pains be genius, then the term rightfully belongs to Pierre Louis Guinand. He was a Swiss artisan living in the Canton of Neuchatel near Chaux-de-Fonds, maker of bells for repeaters, and becoming interested in constructing telescopes imported some flint glass from England and found it bad.

He thereupon undertook the task of making better, and from 1784 kept steadily at his experiments, failure only spurring him on to redoubled efforts. All he could earn at his trade went into his furnaces, until gradually he won success, and his glass began to be heard of; for by 1799 he was producing flawless discs of flint as much as 6 inches in diameter.

What is more, to Guinand is probably due the production of the denser, more highly refractive flints, especially valuable for achromatic telescopes. The making of optical glass has always been an art rather than a science. It is one thing to know the exact composition of a glass and quite another to know in what order and proportion the ingredients went into the furnace, to what temperature they were carried, and for how long, and just how the fused mass must be treated to free the products from bubbles and striæ.

Even today, though much has been learned by scientific investigation in the past few years, it is far from easy to produce two consecutive meltings near enough in refractive power to be treated as optically identical, or to produce large discs optically homogeneous. What Guinand won by sheer experience was invaluable. He was persuaded in 1805 to move to Munich and eventually to join forces with Fraunhofer, an association which made both the German optical glass industry and the modern refractor.

He returned to Switzerland in 1814 and continued to produce perfect discs of larger and larger dimensions. One set of 12 inches worked up by Cauchoix in Paris furnished what was for some years the world's largest refractor.

Guinaud died in 1824, but his son Henry, moving to Paris, brought his treasure of practical knowledge to the glass works there, where it has been handed down, in effect from father to son, gaining steadily by accretion, through successive firms to the present one of Parra-Mantois.

Bontemps, one of the early pupils of Henry Guinand, emigrated to England at the Revolution of 1848 and brought the art to the famous firm of Chance in Birmingham. Most of its early secrets have long been open, but the minute teachings of experience are a tremendously valuable asset even now.

To Fraunhofer, the greatest master of applied optics in the nineteenth century, is due the astronomical telescope in sub-

Fig. 22.—Dr. Joseph von Fraunhofer, the Father of Astrophysics.

stantially its present form. Not only did he become under Guinand's instruction extraordinarily skillful in glass making but he practically devised the art of working it with mathematical precision on an automatic machine, and the science of correctly designing achromatic objectives.

The form which he originated (Fig. 23) was the first in which the aberrations were treated with adequate completeness, and, particularly for small instruments, is unexcelled even now.

The curvatures here shown are extreme, the better to show their relations. The front radius of the crown is about 2½ times longer than the rear radius, the front of the flint is slightly flatter than the back of the crown, and the rear of the flint is only slightly convex.

Fraunhofer's workmanship was of the utmost exactness and it is not putting the case too strongly to say that a first class example of the master's craft, in good condition, would compare well in color-correction, definition, and field, with the best modern instruments.

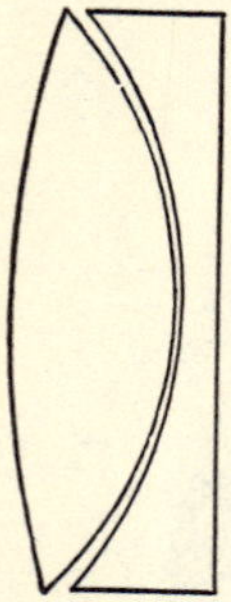

FIG. 23.

The work done by the elder Struve at Dorpat with Fraunhofer's first large telescope (9.6 inches aperture and 170 inches focal length) tells the story of its quality, and the Königsberg heliometer, the first of its class, likewise, while even today some of his smaller instruments are still doing good service.

It was he who put in practice the now general convention of a relative aperture of about F/15, and standardized the terrestrial eyepiece into the design quite widely used today. The improvements since his time have been relatively slight, due mainly to the recent production of varieties of optical glass unknown a century ago. Fraunhofer was born in Straubing, Bavaria, March 6, 1787. Self-educated like Herschel, he attained to an extraordinary combination of theoretical and practical knowledge that went far in laying the foundations of astrophysics.

The first mapping of the solar spectrum, the invention of the diffraction grating and its application to determining the wave length of light, the first exact investigation of the refraction and dispersion of glass and other substances, the invention of the objective prism, and its use in studying the spectra of stars and planets, the recognition of the correspondence of the sodium lines to the D lines in the sun, and the earliest suggestion of the diffraction theory of resolution later worked out by Lord Rayleigh and Professor Abbé, make a long list of notable achievements.

To these may be added his perfecting of the achromatic telescope, the equatorial mounting and its clockwork drive, the improvement of the heliometer, the invention of the stage micrometer, several types of ocular micrometers, and the automatic ruling engine.

He died at the height of his creative powers June 7, 1826, and lies buried at Munich under the sublime ascription, by none better earned, *Approximavit Sidera.*

From Fraunhofer's time, at the hands of Merz his immediate successor, Cauchoix in France, and Tully in England, the achromatic refractor steadily won its way. Reflecting telescopes, despite the sensational work of Lord Rosse on his 6-foot mirror of 53 feet focus (unequalled in aperture until the 6-foot of the Dominion Observatory seventy years later), and the even more successful instrument of Mr. Lassell (4 feet aperture, 39 feet focus), were passing out of use, for the reason already noted, that repolishing meant refiguring and the user had to be at once astronomer and superlatively skilled optician.

These large specula, too, were extremely prone to serious flexure and could hardly have been used at all except for the equilibrating levers devised by Thomas Grubb about 1834, and used effectively on the Rosse instrument. These are in effect a group of upwardly pressing counterbalanced planes distributing among them the downward component of the mirror's weight so as to keep the figure true in any position of the tube.

Such was the situation in the 50's of the last century, when the reflector was quite unexpectedly pushed to the front as a practical instrument by almost simultaneous activity in Germany and France. The starting point in each was Liebig's simple chemical method of silvering glass, which quickly and easily lays on a thin reflecting film capable of a beautiful polish.

The honor of technical priority in its application to silvering telescope specula worked in glass belongs to Dr. Karl August Steinheil (1801–1870) who produced about the beginning of 1856 an instrument of 4-inch aperture reported to have given with a power of 100 a wonderfully good image. The publication was merely from a news item in the "*Allgemeine Zeitung*" of Augsburg, March 24, 1856, so it is little wonder that the invention passed for a time unnoticed.

Early the next year, Feb. 16, 1857, working quite independently, exactly the same thing was brought before the French Academy of Sciences by another distinguished physicist, Jean Bernard Léon Foucault, immortal for his proof of the earth's rotation by

FIG. 24.—Dr. Karl August Steinheil.

FIG. 25.—Jean Bernard Léon Foucault.
The Inventors of the Silver-on-Glass Reflector.

the pendulum experiment, his measurement of the velocity of light, and the discovery of the electrical eddy currents that bear his name.

To Foucault, chiefly, the world owes the development of the modern silver-on-glass reflector, for not being a professional optician he had no hesitation in making public his admirable methods of working and testing, the latter now universally employed. It is worth noting that his method of figuring was,

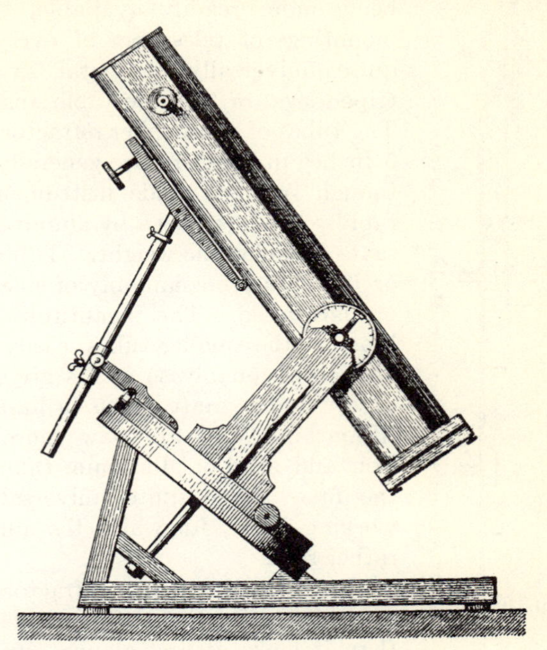

Fig. 26.—Early Foucault Reflector.

physically, exactly what Jesse Ramsden (1735–1800) had pointed out in 1779, (Phil. Tr. 1779, 427) geometrically. One of Foucault's very early instruments mounted equatorially by Sécrétan is shown in Fig. 26.

The immediate result of the admirable work of Steinheil and Foucault was the extensive use of the new reflector, and its rapid development as a convenient and practical instrument, especially in England in the skillful hands of With, Browning, and Calver. Not the least of its advantages was its great superiority over the older type in light-grasp, silver being a better reflector than speculum metal in the ratio of very nearly 7 to 5. From this time on both refractors and reflectors have been fully available to the user of telescopes.

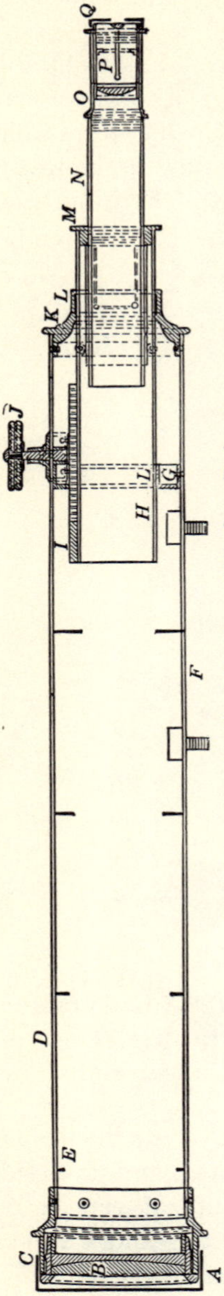

Fig. 27.—Longitudinal Section of Modern Refractor.

In details of construction both have gained somewhat mechanically. As we have seen, tubes were often of wood, and not uncommonly the mountings also. At the present time metal work of every kind being more readily available, tubes and mountings of telescopes of every size are quite universally of metal, save for the tripod-legs of the portable instruments. The tubes of the smaller refractors, say 3 to 5 inches in aperture, are generally of brass, though in high grade instruments this is rapidly being replaced by aluminum, which saves considerable weight. Tubes above 5 or 6 inches are commonly of steel, painted or lacquered. The beautifully polished brass of the smaller tubes, easily damaged and objectionably shiny, is giving way to a serviceable matt finish in hard lacquer. Mountings, too, are now more often in iron and steel or aluminum than in brass, the first named quite universally in the working parts, for which the aluminum is rather soft.

The typical modern refractor, even of modest size, is a good bit more of a machine than it looks at first glance. In principle it is outlined in Fig. 5, in practice it is much more complex in detail and requires the nicest of workmanship. In fact if one were to take completely apart a well-made small refractor, including its optical and mechanical parts one would reckon up some 30 to 40 separate pieces, not counting screws, all of which must be accurately fitted and assembled if the instrument is to work properly.

Fig. 27 shows such an instrument in

section from end to end, as one would find it could he lay it open longitudinally.

A is the objective cap covering the objective *B* in its adjustable cell *C*, which is squared precisely to the axis of the main tube *D*. Looking along this one finds the first of the diaphragms, *E*.

These are commonly 3 to 6 in number spaced about equally down the tube, and are far more important than they look. Their function is not to narrow the beam of light that reaches the ocular, but to trap light which might enter the tube obliquely and be reflected from its sides into the ocular, filling it with stray glare.

No amount of simple blackening will answer the purpose, for even dead black paint such as opticians use reflects at very oblique incidence quite 10 to 20 per cent of the beam. The importance of both diaphragms and thorough blackening has been realized for at least a century and a half, and one can hardly lay too much stress upon the matter

The diaphragms should be so proportioned that, when looking up the tube from the edge of an aperture of just the size and position of the biggest lens in the largest eyepiece, no part of the edge of the objective is cut off, and no part of the side of the tube is visible beyond the nearest diaphragm.

Going further down the tube past a diaphragm or two one comes to the clamping screws *F*. These serve to hold the instrument to its mounting. They may be set in separate bases screwed in place on the inside of the tube, or may be set in the two ends of a lengthwise strap thus secured. They are placed at the balance point as nearly as may be, generally nearer the eye end than the objective.

Then, after one or more diaphragms, comes the guide ring *G*, which steadies the main draw tube *H*, and the rack *I* by which it is moved for the focussing in turning the milled head of the pinion *J*. The end ring *K* of the main tube furnishes the other bearing of *H*, and both *G* and *K* are commonly recessed for accurately fitted cloth lining rings *L*, *L*, to give the draw tube the necessary smoothness of motion.

For the same reason *I* and *J* have to be cut and fitted with the utmost exactness so as to work evenly and without backlash. *H* is fitted at its outer end with a slide ring and tube *M*, generally again cloth lined to steady the sliding eyepiece tube *N*. This is terminated by the spring collar *O*, in which fits the eyepiece *P*, generally of the two lens form; and finally comes the eyepiece cap Q set at the proper distance from the eye lens and with an aperture of carefully determined size.

One thus gets pretty well down in the alphabet without going much into the smaller details of construction. Both objective mount and ocular are somewhat complex in fact, and the former is almost always made adjustable in instruments of above 3 or 4 inches aperture, as shown in Fig. 28, the form used by Cooke, the famous maker of York, England. Unless the optical axis of the objective is true with the tube bad images result.

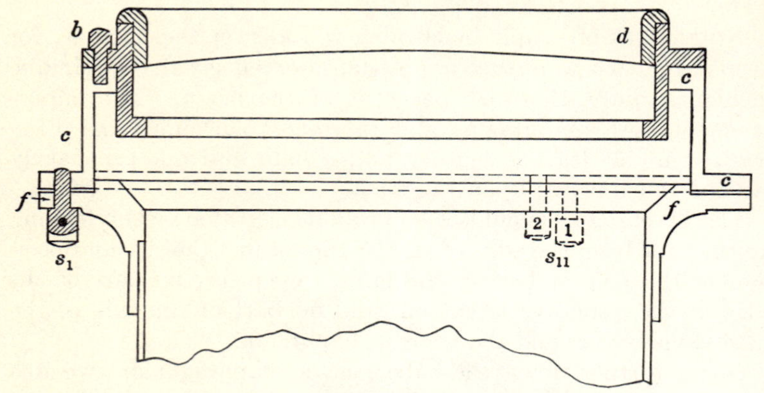

Fig. 28.—Adjustable Cell for Objective.

To the upper end of the tube is fitted a flanged counter-cell c, to an outward flange f, tapped for 3 close pairs of adjusting screws as s_1, s_{11} spaced at 120° apart. The objective cell itself, b, is recessed for the objective which is held in place by an interior or exterior ring d. The two lenses of the achromatic objective are usually very slightly separated by spacers, either tiny bits of tinfoil 120° apart, or a very thin ring with its upper edge cut down save at 3 points.

This precaution is to insure that the lenses are quite uniformly supported instead of touching at uncertain points, and quite usually the pair as a whole rests below on three corresponding spacers. Of each pair of adjusting screws one as 1 in the pair s_{11} is threaded to push the counter cell out, the adjacent one, 2, to pull it in, so that when adjustment is made the objective is firmly held. Of the lenses that form the objective, the concave flint is commonly at the rear and the convex crown in front.

At the eye end the ocular ordinarily consists of two lenses each burnished into a brass screw ring, a tube, flange, cap, and diaphragm arranged as shown in Fig. 29. There are many varieties of ocular as will presently be shown, but this is a typical form. Figure 30 shows a complete modern refractor of four inches aperture on a portable equatorial stand with slow motion in right ascension and diagonal eye piece.

Reflectors, used in this country less than they deserve, are, when properly mounted, likewise possessed of many parts. The smaller ones, such as are likely to come into the reader's hands, are almost always in the Newtonian form, with a small oblique mirror to bring the image outside the tube.

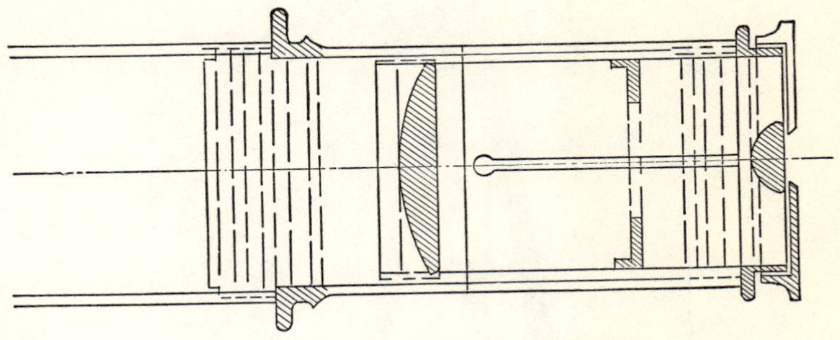

FIG. 29.—The Eyepiece and its Fittings.

The Gregorian form has entirely vanished. Its only special merit was its erect image, which gave it high value as a terrestrial telescope before the days of achromatics, but from its construction it was almost impossible to keep the field from being flooded with stray light, and the achromatic soon displaced it. The Cassegranian construction on the other hand, shorter and with aberrations much reduced, has proved important for obtaining long equivalent focus in a short mount, and is almost universally applied to large reflectors, for which a Newtonian mirror is also generally provided.

Figure 31 shows in section a typical reflector of the Newtonian form. Here *A* is the main tube, fitted near its outer end with a ring *B* carrying the small elliptical mirror *C*, which is set at 45° to the axis of the tube. At the bottom of the tube is the parabolic main mirror *D*, mounted in its cell *E*. Just opposite the 45° small mirror is a hole in the tube to which is fitted the eyepiece mounting F, carrying the eyepiece G, fitted to a spring collar H, screwed into a draw tube I, sliding in its mounting and brought to focus by the rack-and-pinion J.

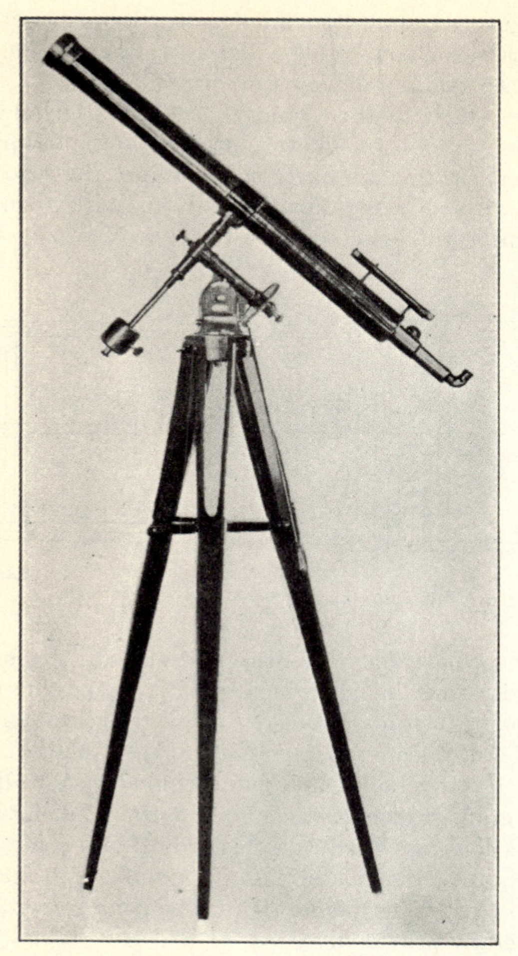

FIG. 30.—Portable Equatorial Refractor (Brashear).

At K, K, are two rings fixed to the tube and bearing smoothly against the rings L L rigidly fixed to the bar M carried by the polar axis of the mount. The whole tube can therefore be rotated about its axis so as to bring the eye piece into a convenient posi-

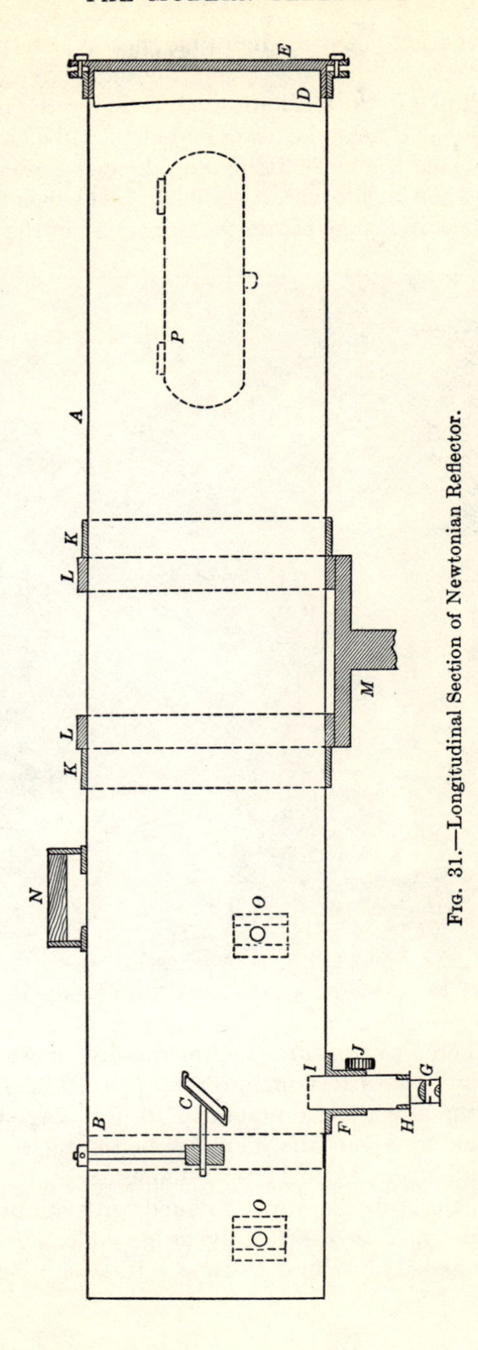

Fig. 31.—Longitudinal Section of Newtonian Reflector.

tion for observation. One or more handles, N, are provided for this purpose.

Brackets shown in dotted lines at O, O, carry the usual finder, and a hinged door P near the lower end of the tube enables one to remove or replace the close fitting metal cover that protects the main mirror when not in use. Similarly a cover is fitted to the small mirror, easily reached from the upper end of the tube. The

FIG. 32.—Reflector with Skeleton Tube (Brashear).

proportions here shown are approximately those commonly found in medium sized instruments, say 7 to 10 inches aperture. The focal ratio is somewhere about $F/6$, the diagonal mirror is inside of focus by about the diameter of the main mirror, and its minor axis is from $\frac{1}{5}$ to $\frac{1}{4}$ that diameter.

Note that the tube is not provided with diaphragms. It is merely blackened as thoroughly as possible, although stray light is quite as serious here as in a refractor. One could fit diaphragms effectively only in a tube of much larger diameter than the mirror, which would be inconvenient in many ways.

A much better way of dealing with the difficulty is shown in Fig. 32 in which the tube is reduced to a skeleton, a construction common in large instruments. Nothing is blacker than a clear opening into the darkness of night, and in addition there can be no localized air currents, which often injure definition in an ordinary tube.

Instruments by different makers vary somewhat in detail. A good type of mirror mounting is that shown in Fig. 33, and used for many years past by Browning, one of the famous English makers. Here the mirror *A*, the back of which is made accurately plane, is seated in its counter-cell *B*, of which a wide annulus *F*, *F*, is also a good plane, and is lightly held in place by a retaining ring. This counter cell rests in the outer cell *C* on three equidistant studs regulated by the concentric push-and-pull adjusting screws *D*, *D*, *E*, *E*. The outer cell may be solid, or a skeleton for lightness and better equalization of temperature.

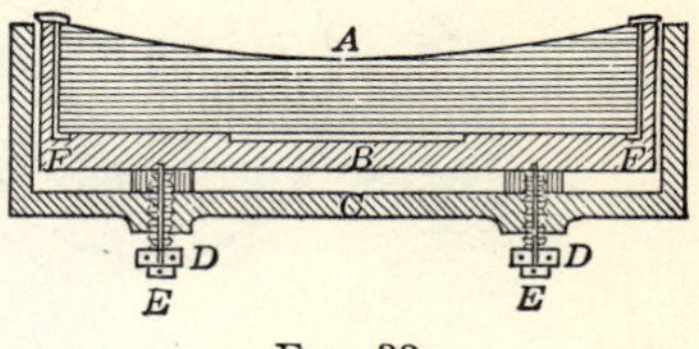

FIG. 33.

Small specula may be well supported on any flat surface substantial enough to be thoroughly rigid, with one or more thicknesses of soft, thick, smooth cloth between, best of all Brussels carpet. Such was the common method of support in instruments of moderate dimensions prior to the day of glass specula. Sir John Herschel speaks of thus carrying specula of more than a hundred-weight, but something akin to Browning's plan is generally preferable.

There is also considerable variety in the means used for supporting the small mirror centrally in the tube. In the early telescopes it was borne by a single stiff arm which was none too stiff and produced by diffraction a long diametral flaring ray in the images of bright stars.

A great improvement was introduced by Browning more than a half century ago, in the support shown in Fig. 34. Here the ring *A*, (*B*, Fig. 31) carries three narrow strips of thin spring steel, *B*, extending radially inward to a central hub which carries the mirror *D*, on adjusting screws *E*. Outside the ring the tension screws C enable the mirror to be accurately centered and held in place. Rarely, the mirror is replaced by a totally reflecting right angled prism which saves some light, but unless for small instruments is rather heavy and hard to obtain of the requisite quality and precision of figure. A typical modern reflector by Brashear, of 6 inches aperture, is shown in Fig. 35, complete with circles and driving clock, the latter contained in the hollow iron pier, an arrangement usual in American-made instruments.

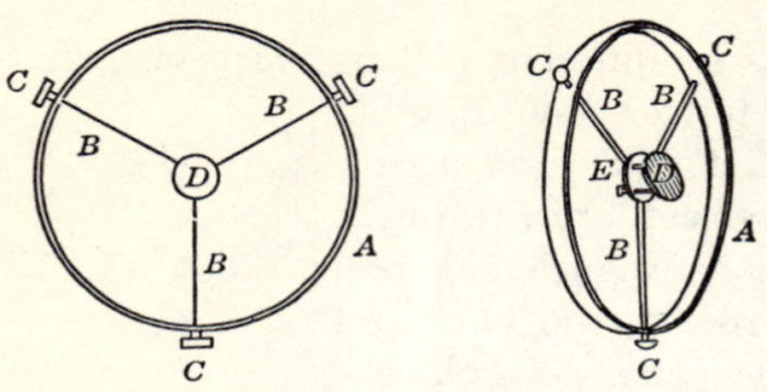

FIG. 34.—Support of Diagonal Mirror (Browning.)

Recent reflectors, particularly in this country, have four supporting strips instead of three, which gives a little added stiffness, and produces in star images but four diffraction rays instead of the six produced by the three strip arrangement, each strip giving a diametral ray.

In some constructions the ring A is arranged to carry the eyepiece fittings, placed at the very end of the tube and arranged for rotating about the optical axis of the telescope. This allows the ocular to be brought to any position without turning the whole tube. In small instruments a fixed eyepiece can be used without much inconvenience if located on the north side of the tube (in moderate north latitudes).

Reflectors are easily given a much greater relative aperture than is practicable in a single achromatic objective. In fact they are usually given apertures of $F/5$ to $F/8$ and now and then are pushed to or even below $F/3$. Such mirrors have been successfully used for photography;[1] and less frequently for visual observation, mounted in the Cassegrainian form, which commonly increases the equivalent focal length at least three or four times. A telescope so arranged, with an aperture of a foot or more as in

[1] An $F/3$ mirror of $1m$ aperture by Zeiss was installed in the observatory at Bergedorf in 1911, and a similar one by Schaer is mounted at Carre, near Geneva.

some recent examples, makes a very powerful and compact instrument.

This is the form commonly adopted for the large reflectors of

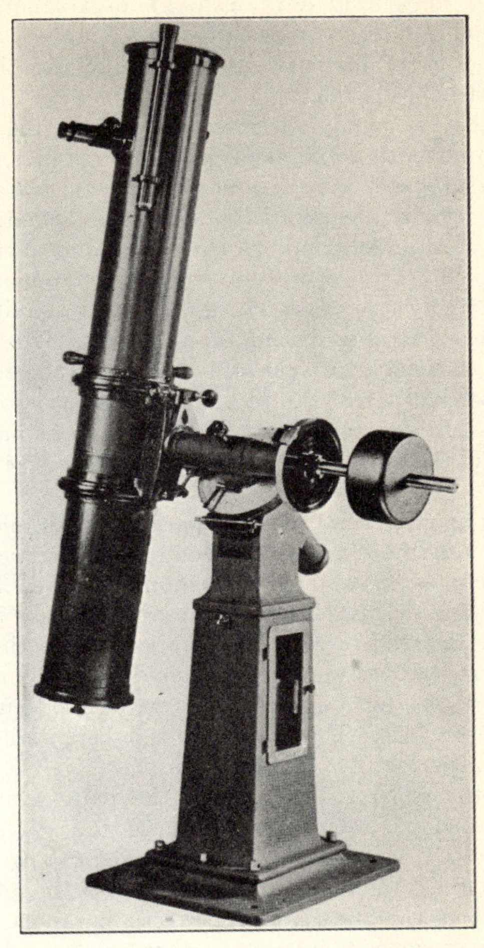

Fig. 35.—Small Equatorially Mounted Reflector.

recent construction, a type being the 60-inch telescope of the Mount Wilson Observatory of which the primary focus is 25¼ feet and the ordinary equivalent focus as a Cassegranian 80 feet.

Comparatively few small reflectors have been made or used in the United States, although the climatic conditions here are more favorable than in England, where the reflector originated and has been very fully developed. The explanation may lie in our smaller number of non-professional active astronomers who are steadily at observational work, and can therefore use reflectors to the best advantage.

The relative advantages of refractors and reflectors have long been a matter of acrimonious dispute. In fact, more of the genuine *odium theologicum* has gone into the consideration of this matter than usually attaches to differences in scientific opinion. A good many misunderstandings have been due to the fact that until recently few observers were practically familiar with both instruments, and the professional astronomer was a little inclined to look on the reflector as fit only for amateurs. The comparison is somewhat clarified at present by the fact that the old speculum metal reflector has passed out of use, and the case now stands as between the ordinary refracting telescope such as has just been described, and the silver-on-glass reflector discussed immediately thereafter.

The facts in the case are comparatively simple. Of two telescopes having the same clear aperture, one a reflector and the other a refractor, each assumed to be thoroughly well figured, as it can be in fact today, the theoretical resolving power is the same, for this is determined merely by the aperture, so that the only possible difference between the two would be in the residual imperfection in the performance of the refractor due to its not being perfectly achromatic. This difference is substantially a negligible one for many, but not all, purposes.

Likewise, the general definition of the pair, assuming first-class workmanship, would be equal. Of the two, the single surface of the mirror is somewhat more difficult to figure with the necessary precision than is any single surface of the refractor, but reflectors can be, and are, given so perfect a parabolic figure that the image is in no wise inferior to that produced by the best refractors, and the two types of telescopes will stand under favorable circumstances the same proportional magnifying powers.

The mirror is much more seriously affected by changes of temperature and by flexure than is the objective, since in the former case the successive surfaces of the two lenses in the achromatic combination to a considerable extent compensate each other's slight changes of curvature, which act only by still slighter changes of refraction, while the mirror surface stands alone and any change in curvature produces double the defect on the reflected ray.

It is therefore necessary, as we shall see presently, to take particular precautions in working with a reflecting telescope, which is, so to speak, materially more tender as regards external conditions than the refractor. As regards light-grasp, the power of rendering faint objects visible, there is more room for honest variety of opinion. It was often assumed in earlier days that a reflector was not much brighter than a refractor of half the aperture, *i.e.*, of one quarter the working area.

This might have been true in the case of an old speculum metal reflector in bad condition, but is certainly a libel on the silver-on-glass instrument, which Foucault on the other hand claimed to be, aperture for aperture, brighter than the refractor. Such a relation might in fact temporarily exist, but it is far from typical.

The real relation depends merely on the light losses demonstrably occurring in the two types of telescopes. These are now quite well known. The losses in a refractor are those due to absorption of light in the two lenses, plus those due to the four free surfaces of these lenses. The former item in objectives of moderate size aggregates hardly more than 2 to 3 per cent. The latter, assuming the polish to be quite perfect, amount to 18 to 20 per cent of the incident light, for the glasses commonly used.

The total light transmitted is therefore not over 80 per cent of the whole, more often somewhat under this figure. For example, a test by Steinheil of one of Fraunhofer's refractors gave a transmission of 78 per cent, and other tests show similar results.

The relation between the light transmitted by glass of various thickness is very simple. If unit thickness transmits m per cent of the incident light then n units in thickness will pass m^n per cent. Thus if one half inch passes .98, two inches will transmit $.98^4$, or .922. Evidently the bigger the objective the greater the absorptive loss. If the loss by reflection at a single surface leaves m per cent to be transmitted then n surfaces will transmit m^n. And m being usually about .95, the four surfaces of an objective let pass nearly .815, and the thicker objective as a whole transmits approximately 75 per cent.

As to the reflector the whole relation hinges on the coefficient of reflection from a silvered surface, under the circumstances of the comparison.

In the case of a reflecting telescope as a whole, there are commonly two reflections from silver and if the coefficient of reflection is m then the total light reflected is m^2. Now the reflectivity of a silver-on-glass film has been repeatedly measured. Chant (Ap. J. **21,** 211) found values slightly in excess of 95 per cent, Rayleigh (Sci. Papers **2, 4**) got 93.9, Zeiss (Landolt u. Bornstein, Tabellen) about 93.0 for light of average wave length.

Taking the last named value, a double reflection would return substantially 86.5 per cent of the incident light. No allowance is here made for any effect of selective reflection, since for the bright visual rays, which alone we are considering, there is very slight selective effect. In the photographic case it must be taken into account, and the absorption in glass becomes a serious factor in the comparison, amounting for the photographic rays to as much as 30 to 40 per cent in large instruments. Now in comparing reflector and refractor one must subtract the light stopped by the small mirror and its supports, commonly from 5 to 7 per cent. One is therefore forced to the conclusion that with silver coatings fresh and very carefully polished reflector and refractor will show for equal aperture equal light grasp.

But as things actually go even fresh silver films are quite often below .90 in reflectivity and in general tarnish rather rapidly, so that in fact the reflector falls below the refractor by just about the amount by which the silver films are out of condition. For example Chant (loc. cit.) found after three months his reflectivity had fallen to .69. A mirror very badly tarnished by fifteen weeks of exposure to dampness and dust, uncovered, was found by the writer down to a scant .40.

The line of Fig. 36 shows the relative equivalent apertures of refractors corresponding to a 10-inch reflector at coefficients of reflection for a single silvered surface varying from .95 to .50 at which point the film would be so evidently bad as to require immediate renewal. The relation is obviously linear when the transmission of the objective is, as here, assumed constant. The estimates of skilled observers from actual comparisons fall in well with the line, showing reflectivities generally around .80 to .85 for well polished films in good condition.

The long and short of the situation is that a silvered reflector deteriorates and at intervals varying from a few months to a year or two depending on situation, climate, and usage, requires repolishing or replacement of the film. This is a fussy job, but quickly done if everything goes well.

As to working field the reflector as ordinarily proportioned is at a disadvantage chiefly because it works at $F/5$ or $F/6$ instead of

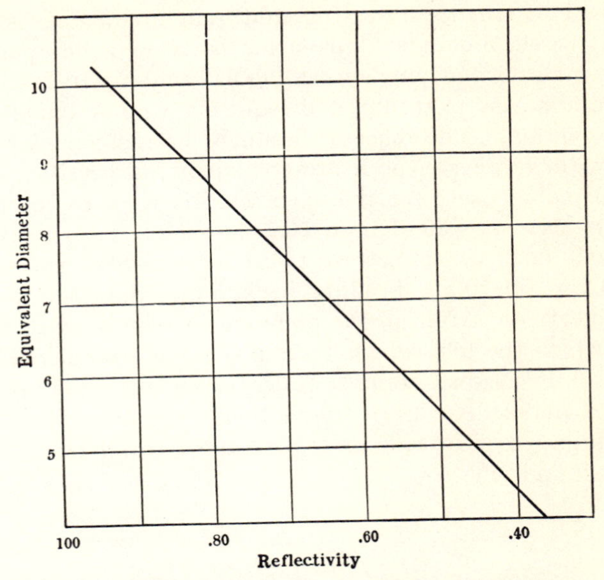

FIG. 36.—Relative Light-grasp of Reflector and Refractor.

at $F/15$. At equal focal ratios there is no substantial difference between reflector and refractor in this respect, unless one goes into special constructions, as in photographic telescopes.

In two items, first cost and convenience in observing, the reflector has the advantage in the moderate sizes. Roughly, the reflector simply mounted costs about one half to a quarter the refractor of equal light grasp and somewhat less resolving power, the discrepancy getting bigger in large instruments (2 feet aperture and upwards).

As to ease of observing, the small refractor is a truly neck-wringing instrument for altitudes above 45° or thereabouts, just the situation in which the equivalent reflector is most convenient. In considering the subject of mounts these relations will appear more clearly.

Practically the man who is observing rather steadily and can give his telescope a fixed mount can make admirable use of a reflector and will not find the perhaps yearly or even half yearly re-silvering at all burdensome after he has acquired the knack—chiefly cleanliness and attention to detail.

If, like many really enthusiastic amateurs, he can get only an occasional evening for observing, and from circumstances has to use a portable mount set up on his lawn, or even roof, when fortune favors an evening's work, he will find a refractor always in condition, easy to set up, and requiring a minimum of time to get into action. The reflector is much the more tender instrument, with, however, the invaluable quality of precise achromatism, to compensate for the extra care it requires for its best performance. It suffers more than the refractor, as a rule, from scattered light, for imperfect polish of the film gives a field generally presenting a brighter background than the field of a good objective. After all the preference depends greatly on the use to which the telescope is to be put. For astrophysical work in general, Professor George E. Hale, than whom certainly no one is better qualified to judge, emphatically endorses the reflector. Most large observatories are now-a-days equipped with both refractors and reflectors.

CHAPTER III

OPTICAL GLASS AND ITS WORKING

Glass, one of the most remarkable and useful products of man's devising, had an origin now quite lost in the mists of antiquity. It dates back certainly near a thousand years before the Christian era, perhaps many centuries more. Respecting its origin there are only traditions of the place, quite probably Syria, and of the accidental melting together of sand and soda. The product, sodium silicate, readily becomes a liquid, i.e., "water-glass," but the elder Pliny, who tells the story, recounts the later production of a stable vitreous body by the addition of a mineral which was probably a magnesia limestone.

This combination would give a good permanent glass, whether the story is true or not, and very long before Pliny's time glass was made in great variety of composition and color. In fact in default of porcelain, glass was used in Roman times relatively more than now. But without knowledge of optics there was no need for glass of optical quality; it was well into the Renaissance before its manufacture had reached a point where anything of the sort could be made available even in small pieces, and it is barely over a century since glass-making passed beyond the crudest empiricism.

Glass is substantially a solid solution of silica with a variety of metallic oxides, chiefly those of sodium, potassium, calcium and lead, sometimes magnesium, boron, zinc, barium and others.

By itself silica is too refractory to work easily, though silica glass has some very valuable properties, and the alkaline oxides in particular serve as the fluxes in common use. Other oxides are added to obtain various desired properties, and some impurities may go with them.

The melted mixture is thus a somewhat complex solution containing frequently half a dozen ingredients. Each has its own natural melting and vaporizing point, so that while the blend remains fairly uniform it may tend to lose some constituent while molten, or in cooling to promote the crystallization of another, if held too near its particular freezing point. Some combinations are more likely to give trouble from this cause than others, and while a very wide variety of oxides can be coerced into solution with silica, a comparatively limited number produce a homogeneous and colorless glass useful for optical purposes.

Many mixtures entirely suitable for common commercial purposes are out of the question for lens making, through tendency to surface deterioration by weathering, lack of homogeneous quality, or objectionable coloration. A very small amount of iron in the sand used at the start gives the green tinge familiar in cheap bottles, which materially decreases the transparency. The bottle maker often adds oxide of manganese to the mixture, which naturally of itself gives the glass a pinkish tinge, and so apparently whitens it by compensating the one absorption by another. The resulting glass looks all right on a casual glance, but really cuts off a very considerable amount of light.

A further difficulty is that glass differs very much in its degree of fluidity, and its components sometimes seem to undergo mutual reactions that evolve persistent fine bubbles, besides reacting with the fireclay of the melting pot and absorbing impurities from it.

The molten glass is somewhat viscous and far from homogeneous. Its character suggests thick syrup poured into water, and producing streaks and eddies of varying density. Imagine such a mixture suddenly frozen, and you have a good idea of a common condition in glass, transparent, but full of striæ. These are frequent enough in poor window glass, and are almost impossible completely to get rid of, especially in optical glass of some of the most valuable varieties.

The great improvement introduced by Guinand was constant stirring of the molten mass with a cylinder of fire clay, bringing bubbles to the surface and keeping the mass throughly mixed from its complete fusion until, very slowly cooling, it became too viscous to stir longer.

The fine art of the process seems to be the exact combination of temperature, time, and stirring, suitable for each composition of the glass. There are, too, losses by volatilization during melting, and even afterwards, that must be reckoned with in the proportions of the various materials put into the melting, and in the temperatures reached and maintained.

One cannot deduce accurately the percentage mixture of the raw materials from an analysis of the glass, and it is notorious that the product even of the best manufacturers not infrequently fails to run quite true to type. Therefore the optical properties of each melting have carefully to be ascertained, and the product listed either as a very slight variant from its standard type, or as an odd lot, useful, but quite special in properties. Some of these odd meltings in fact have optical peculiarities the regular reproduction of which would be very desirable.

The purity of the materials is of the utmost importance in producing high grade glass for optical or other purposes. The silica is usually introduced in the form of the purest of white sand carrying only a few hundredths of one per cent of impurities in the way of iron, alumina and alkali. The ordinary alkalis go in preferably as carbonates, which can be obtained of great purity; although in most commercial glass the soda is used in the form of "salt-cake," crude sodium sulphate.

Calcium, magnesium, and barium generally enter the melt as carbonates, zinc and lead as oxides. Alumina, like iron, is generally an impurity derived from felspar in the sand, but occasionally enters intentionally as pure natural felspar, or as chemically prepared hydrate. A few glasses contain a minute amount of arsenic, generally used in the form of arsenious acid, and still more rarely other elements enter, ordinarily as oxides.

Whatever the materials, they are commonly rather fine ground and very thoroughly mixed, preferably by machinery, before going into the furnaces. Glass furnaces are in these days commonly gas fired, and fall into two general classes, those in which the charge is melted in a huge tank above which the gas flames play, and those in which the charge is placed in crucibles or pots open or nearly closed, directly heated by the gas. In the tank furnaces the production is substantially continuous, the active melting taking place at one end, where the materials are introduced, while the clear molten glass flows to the cooler end of the tank or to a cooler compartment, whence it is withdrawn for working.

The ordinary method of making optical glass is by a modification of the pot process, each pot being fired separately to permit better regulation of the temperature.

The pots themselves are of the purest of fire clay, of moderate capacity, half a ton or so, and arched over to protect the contents from the direct play of the gases, leaving a side opening sufficient for charging and stirring.

The fundamental difference between the making of optical glass and the ordinary commercial varieties lies in the individual treatment of each charge necessary to secure uniformity and regularity, carried even to the extent of cooling each melting very slowly in its own pot, which is finally broken up to recover the contents. The tank furnaces are under heat week in and week out, may hold several hundred tons, and on this account cannot so readily be held to exactness of composition and quality.

The optical glass works, too, is provided with a particularly efficient set of preheating and annealing kilns, for the heat treatment of pots and glass must be of the most careful and thorough kind.

The production of a melting of optical glass begins with a very gradual heating of the pot to a bright red heat in one of the kilns. It is then transferred to its furnace which has been brought to a similar temperature, sealed in by slabs of firebrick, leaving its mouth easy of access, and then the heat is pushed up to near the melting temperature of the mixture in production, which varies over a rather wide range, from a moderate white heat to the utmost that a regenerative gas furnace can conveniently produce. After the heating comes the rather careful process of charging.

The mixture is added a portion at a time, since the fused material tends to foam, and the raw material as a solid is more bulky than the fluid. The chemical reactions as the mass fuses are somewhat complex. In their simplest form they represent the formation of silicates.

At high temperatures the silica acts as a fairly strong acid, and decomposes the fused carbonates of sodium and potassium with evolution of gas. This is the *rationale* of the fluxing action of such alkaline substances of rather low melting point. Other mixtures act somewhat analogously but in a fashion commonly too complex to follow.

The final result is a thick solution, and the chief concern of the optical glass maker is to keep it homogeneous, free from bubbles, and as nearly colorless as practicable. To the first two ends the temperature is pushed up to gain fluidity, and frequently substances are added (e.g., arsenic) which by volatility or chemical effect tend to form large bubbles from the entrained gases, capable of clearing themselves from the fluid where fine bubbles would remain. For the same purpose is the stirring process.

The stirrer is a hard baked cylinder of fire clay fastened to an iron bar. First heated in the mouth of the pot, the stirrer is plunged in the molten glass and given a steady rotating motion, the long bar being swivelled and furnished with a wooden handle for the workman. This stirring is kept up pretty steadily while the heat is very slowly reduced until the mass is too thick to manage, the process taking, for various mixtures and conditions, from three or four hours to the better part of a day.

Then begins the careful and tedious process of cooling. Fairly rapid until the mass is solid enough to prevent the formation of fresh striæ, the cooling is continued more slowly, in the furnace

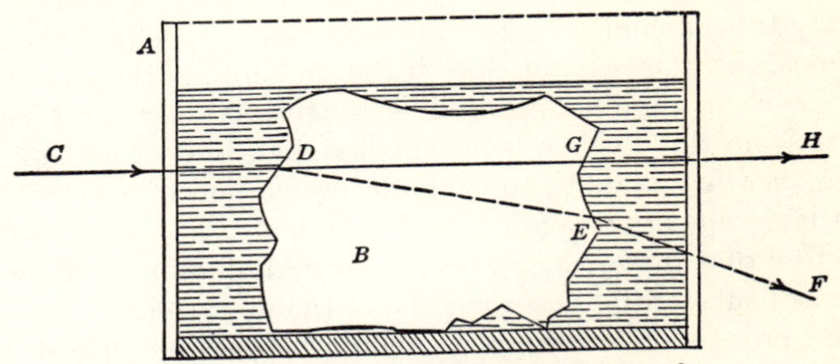

FIG. 37.—Testing Optical Glass in the Rough.

or after removal to the annealing oven, until the crucible is cool enough for handling, the whole process generally taking a week or more.

Then the real trouble begins. The crucible is broken away and there is found a more or less cracked mass of glass, sometimes badly broken up, again furnishing a clear lump weighing some hundreds of pounds. This glass is then carefully picked over and examined for flaws, striæ and other imperfections.

These can sometimes be chipped away with more or less breaking up of the mass. The inspection of the glass in the raw is facilitated by the scheme shown in elevation Fig. 37. Here A is a tank with parallel sides of plate glass. In it is placed B the rough block of glass, and the tank is then filled with a liquid which can be brought to the same refractive power as the glass, as in Newton's disastrous experiment. When equality is reached for, say, yellow light, one can see directly through the block, the rays no longer being refracted at its surface, and any interior striæ are readily seen even in a mass a foot or more thick. Before adding the liquid a ray would be skewed, as *C*, *D*, *E*, *F*, afterwards it would go straight through; *C*, *D*, *G*, *H*.

The fraction that passes inspection may be found to be from much less than a quarter to a half of the whole. This good glass is then ready for the next operation, forming and fine annealing. The final form to be reached is a disc or block, and the chunks of perfect glass are heated in a kiln until plastic, and then moulded into the required shapes, sometimes concave or convex discs suitable for small lenses.

Then the blocks are transferred to a kiln and allowed to cool off very gradually, for several days or weeks according to the size of the blocks and the severity of the requirements they must meet. In the highest class of work the annealing oven has thermostatic control and close watch is kept by the pyrometer.

It is clear that the chance of getting a large and perfect chunk from the crucible is far smaller than that of getting fragments of a few pounds, so that the production of a perfect disc for a large objective requires both skill and luck. Little wonder therefore that the price of discs for the manufacture of objectives increases substantially as the cube of the diameter.

The process of optical glass-making as here described is the customary one, used little changed since the days of Guinand. The great advances of the last quarter century have been in the production of new varieties having certain desirable qualities, and in a better understanding of the conditions that bring a uniform product of high quality. During the world war the greatly increased demand brought most extraordinary activity in the manufacture, and especially in the scientific study of the problems involved, both here and abroad. The result has been a long step toward quantity production, the discovery that modifications of the tank process could serve to produce certain varieties of optical glass of at least fair quality, and great improvements in the precision and rapidity of annealing.

These last are due to the use of the electric furnace, the study of the strains during annealing under polarized light, and scientific pyrometry. It is found that cooling can be much hastened over certain ranges of temperature, and the total time required very greatly shortened. It has also been discovered, thanks to captured instruments, that some of the glasses commonly regarded as almost impossible to free from bubbles have in fact yielded to improved methods of treatment.

Conventionally optical glass is of two classes, crown and flint. Originally the former was a simple compound of silica with soda and potash, sometimes also lime or magnesia, while the latter was rich in lead oxide and with less of alkali. The crown had a low index of refraction and small dispersion, the flint a high index and strong dispersion. Crown glass was the material of general use, while the flint glass was the variety used in cut glass manufacture by reason of its brilliancy due to the qualities just noted.

The refractive index is the ratio between the sine of the angle of incidence on a lens surface and that of the angle of refraction in passing the surface. Fig. 38 shows the relation of the inci-

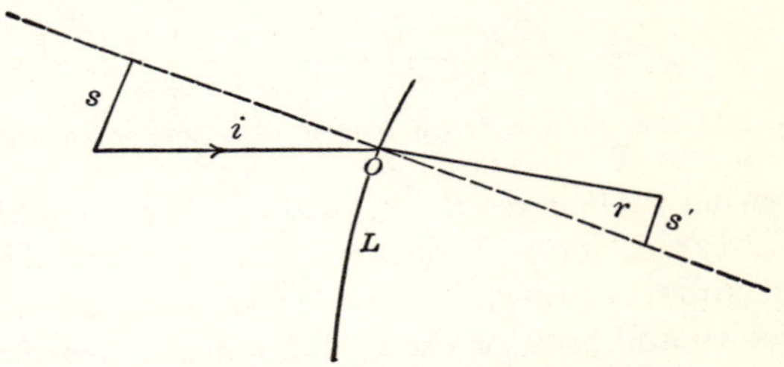

FIG. 38.—The Index of Refraction.

dent and refracted rays in passing from air into the glass lens surface L, and the sines of the angles which determine n, the conventional symbol for the index of refraction. Here i is the angle of incidence and r the angle of refraction, i.e. $n = \frac{s}{s'}$. The indices of refraction are usually given for specific colors representing certain lines in the spectrum, commonly A^1, the potassium line in the extreme red, C the red line due to hydrogen, D the sodium line, F the blue hydrogen line and G' the blue-violet line hydrogen line, and are distinguished as n_c, n_d, n_f, etc. The standard dispersion (dn) for visual rays is given as between C and F, while the standard refractivity is taken for D, in the bright yellow part of the spectrum. (Note. For the convenience of those who are rusty on their trigonometry, Fig. 39 shows the simpler trigonometric functions of an angle. Thus the sine of the angle A is, numerically, the length of the radius divided into the length of the line dropped from the end of the radius to the horizontal base line, i.e. $\frac{bc}{Ob}$, the tangent is $\frac{da}{Ob}$, and the cosine $\frac{Oc}{Ob}$.

Ordinarily the index of refraction of the crown was taken as about $3/2$, that of the flint as about $8/5$. As time has gone on and especially since the new glasses from the Jena works were introduced about 35 years ago, one cannot define crowns and flints in any such simple fashion, for there are crowns of high index and flints of low dispersion.

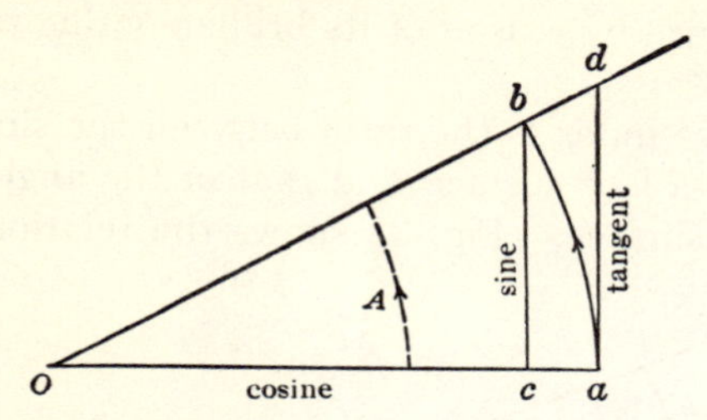

FIG. 39.—The Simple Trigonometric Functions of an Angle.

The following table gives the optical data and chemical analyses of a few typical optical glasses. The list includes common crowns and flints, a typical baryta crown and light flint, and a telescope crown and flint for the better achromatization of objectives, as developed at the Jena works.

The thing most conspicuous here as distinguishing crowns from flints is that the latter have greater relative dispersion in the blue, the former in the red end of the spectrum, as shown by the bracketed ratios. This as we shall see is of serious consequence in making achromatic objectives. In general, too, the values of ν for flints are much lower than for crowns, and the indices of refraction themselves commonly higher.

As we have just seen, glass comes to the optician in blocks or discs, for miscellaneous use the former, three or four inches square and an inch think, more or less; for telescope making the latter. The discs are commonly some ten percent greater in diameter than the finished objective for which they are intended, and in thickness from $1/8$ to $1/10$ the diameter. They are commonly well annealed and given a preliminary polish on both sides to facilitate close inspection.

The first step toward the telescope is the testing of these discs of glass, first for the presence or absence of striæ and other

CHARACTERISTICS OF OPTICAL GLASSES

Glass	n_d	$\frac{dn}{(F—C)}$	ν	Bracketed numbers are proportions of dn			Analysis of glasses in percentages														
				$\frac{D—A'}{dn}$	$\frac{F—D}{dn}$	$\frac{G'—F}{dn}$	SiO_2	B_2O_3	ZnO	PbO	BaO	K_2O	Na_2O	CaO	Al_2O_3	As_2O_5	As_2O_3	Fe_2O_3	Mn_2O_3	Sb_2O_3	MgO
Boro-silicate crown....	1.5069	.00813	62.3	.00529 (.651)	.00569 (.701)	.00457 (.562)	74.8	5.9	...	....	....	7.11	11.3	...	.75	...	.06	...	.06		
Zinco-silicate (hard) crown.	1.5170	.00859	60.2	.00555 (.646)	.00605 (.704)	.00485 (.565)	65.4	2.5	2.0	....	9.6	15.0	5.0	...	...	...	.4	...	.1		
Dense baryta crown...	1.5899	.00970	60.8	.00621 (.640)	.00683 (.704)	.00546 (.563)	37.5	15.0	...	...	41.0	...	....		5.0	1.5					
Baryta light flint......	1.5718	.01133	50.4	.00706 (.623)	.00803 (.709)	.00660 (.582)	51.7	....	7.0	10.0	20.0	9.5	1.5	...	....	.30					
Common light flint....	1.5710	.01327	43.0	.00819 (.617)	.00943 (.710)	.00791 (.596)	54.3	1.5	...	33.0	....	8.0	3.0	...	....	.20					
Common dense flint...	1.6116	.01638	37.3	.00995 (.607)	.01170 (.714)	.00991 (.607)	54.8	....	....	37.0	....	5.8	.8	.60	.4	...	...	.70	...	....	.20
Very dense flint.......	1.6489	.01919	33.8	.01152 (.600)	.01372 (.714)	.01180 (.615)	40.0	....	...	52.6	....	6.5	.5	...	...	.30	...	...	.09		
Densest flint.........	1.7541	.02743	27.5	.01607 (.585)	.01974 (.720)	.01730 (.630)	29.3	....	...	67.5	....	3.0	....	...	...	...	.20	...	.04		
[1] Telescope crown.....	1.5285	.00866	61.0	.00557 (.643)	.00610 (.705)	.00493 (.570)	55.2	....	...	....	22.0	5.7	7.5	5.9	...	...	...	3.7			
[1] Telescope flint......	1.5286	.01025	51.6	.00654 (.638)	.00723 (.705)	.00591 (.576)	59.9	12.7	...	....	....	5.1	3.5	...	...	...	...	2.7		16.1	

[1] Optical data close approximations only.

imperfections; second, for the perfection of the annealing. The maker has usually looked out for all the grosser imperfections before the discs left his works, but a much closer inspection is needed in order to make the best use of the glass.

Bad striæ are of course seen easily, as they would be in a window pane, but such gross imperfections are often in reality less damaging than the apparently slighter ones which must be searched for. The simplest test is to focus a good telescope on

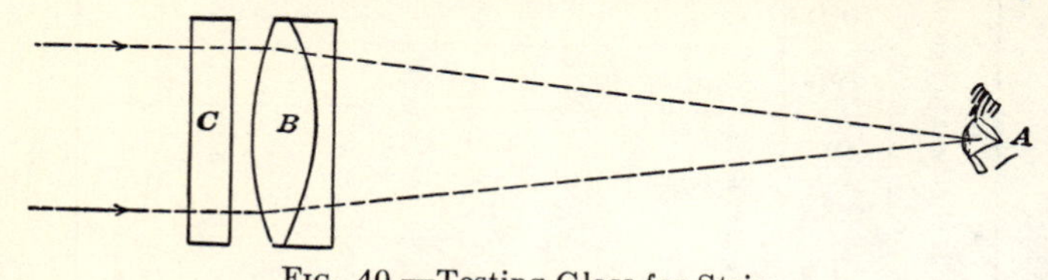

FIG. 40.—Testing Glass for Striæ.

an artificial star, remove the eyepiece and bring the eye into its place.

When the eye is in focus the whole aperture of the objective is uniformly filled with light, and if the disc to be tested be placed in front of it, any inequality in refraction will announce itself by an inequality of illumination. A rough judgment as to the seriousness of the defect may be formed from the area affected and the amount by which it affects the local intensity of illumination. Fig. 40 shows the arrangement for the test, *A* being the eye, *B* the

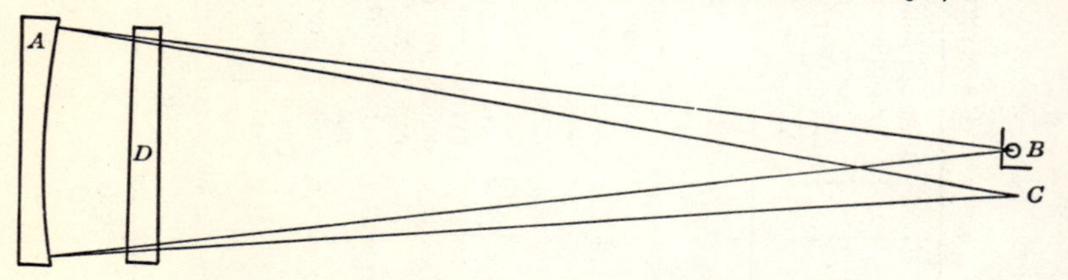

FIG. 41.—The Mirror Test for Striæ.

objective and *C* the disc. The artificial star is conveniently made by setting a black bottle in the sun a hundred feet or so away and, getting the reflection from its shoulder.

A somewhat more delicate test, very commonly used, is shown in Fig. 41. Here *A* is a truly spherical mirror silvered on the front. At *B* very close to its centre of curvature is placed a lamp with a screen in front of it perforated with a hole 1⁄32 inch or so in diameter.

The rays reflected from the mirror come back quite exactly upon themselves and when the eye is placed at *C*, their reflected focus, the whole mirror *A* is uniformly lighted just as the lens was in Fig. 40, with the incidental advantage that it is much easier and cheaper to obtain a spherical mirror for testing a sizeable disc than an objective of similar size and quality. Now placing the disc *D* in front of the mirror, the light passing twice through it shows up the slightest stria or other imperfection as a streak or spot in the field. Its place is obvious and can be at once marked on the glass, but its exact position in the substance of the disc is not so obvious.

To determine this, which may indicate that the fault can be ground out in shaping the lens, a modification of the first test serves well, as indeed it does for the general examination of large discs. Instead of using a distant artificial star and a telescope,

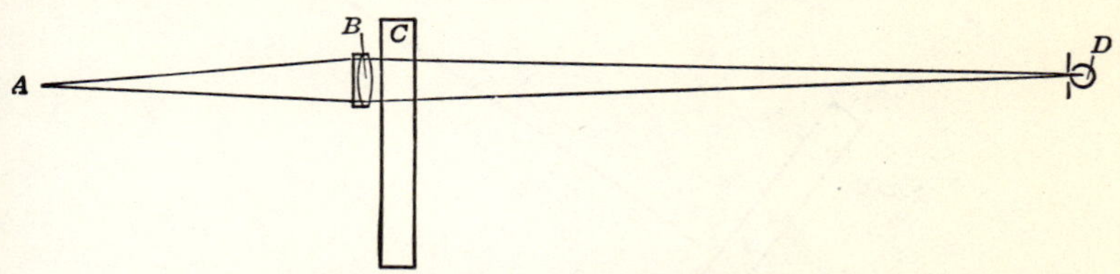

Fig. 42.—Locating Striæ in the Substance of a Disc.

one uses the lamp and screen, or even a candle flame ten feet or more away and a condensing lens of rather short focus, which may or may not be achromatic, so that the eye will get into its focus conveniently while the lens is held in the hand. Fig. 42 shows the arrangement. Here *A* is the eye, *B* the condensing lens, *C* the disc and *D* the source of light. The condensing lens may be held on either side of the disc as convenience suggests, and either disc or lens may be moved. The operation is substantially the examination of a large disc piece-meal, instead of all at once by the use of a big objective or mirror.

Now when a stria has been noted mark its location as to the surface, and, moving the eye a little, look for parallax of the fault with respect to the surface mark. If it appears to shift try a mark on the opposite surface in the same way. Comparison of the two inspections will show about where the fault lies with respect to the surfaces, and therefore what is the chance of working it out. Sometimes a look edgewise of the disc will help in the diagnosis.

Numerous barely detectable striæ are usually worse than one or two conspicuous ones, for the latter frequently throw the light they transmit so wide of the focus that it does not affect the image, which could be greatly damaged by slight blurs of light that just miss focus.

Given a disc that passes well the tests for striæ and the like the next step is to examine the perfection of the annealing, which in its larger aspect is revealed by an examination in polarized light.

For this purpose the disc is set up against a frame placed on table or floor with a good exposure to skylight behind it, and

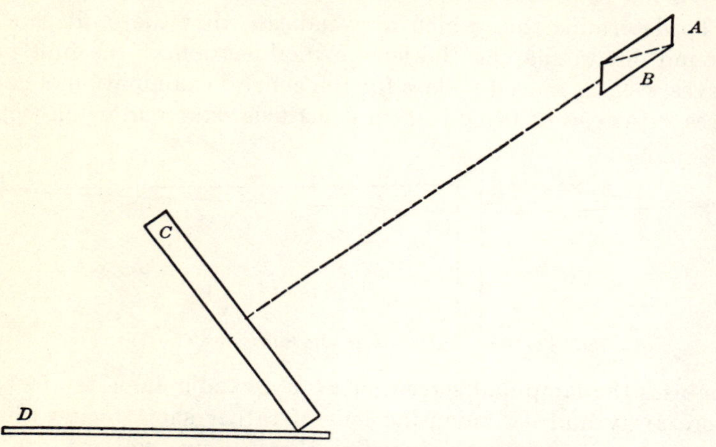

FIG. 43.—Testing a Disc in Polarized Light.

inclined about 35° from the vertical. Behind it is laid a flat shiny surface to serve as polarizer. Black enamel cloth smoothly laid, a glass plate backed with black paint, or even a smooth board painted with asphalt paint will answer excellently. Then holding a Nicol prism before the eye and looking perpendicular to the face of the disc, rotate the prism on its axis. Fig. 43 shows the arrangement, *A* being the eye, *B* the Nicol, *C* the disc, and *D* the polarizer behind it.

If annealing has left no strain the only effect of rotating the Nicol will be to change the field from bright to dark and back again as if the disc were not there. Generally a pattern in the form of a somewhat hazy Maltese cross will appear, with its arms crossing the disc, growing darker and lighter alternately as the Nicol is turned.

If the cross is strongly marked but symmetrical and well centered the annealing is fair—better as the cross is fainter and hazier—altogether bad if colors show plainly or if the cross is decentered or distorted. The test is extremely sensitive, so that holding a finger on the surface of the disc may produce local strain that will show as a faint cloudy spot.

A disc free of striæ and noticeable annealing strains is usually, but not invariably, good, for too frequent reheating in the moulding or annealing process occasionally leaves the glass slightly altered, the effect extending, at worst, to the crystallization or devitrification to which reference has been made.

Given a good pair of discs the first step towards fashioning them into an objective is roughing to the approximate form desired. As a guide to the shaping of the necessary curves, templets must be made from the designed curves of the objective as precisely as possible. These are laid out by striking the necessary radii with beam compass or pivoted wire and scribing the curve on thin steel, brass, zinc or glass. The two last are the easier to work since they break closely to form.

From these templets the roughing tools are turned up, commonly from cast iron, and with these, supplied with carborundum or even sand, and water, the discs, bearing against the revolving tool, are ground to the general shape required. They are then secured to a slowly revolving table, bearing edgewise against a revolving grindstone, and ground truly circular and of the proper final diameter.

At this point begins the really careful work of fine grinding, which must bring the lens very close to its exact final shape. Here again tools of cast iron, or sometimes brass, are used, very precisely brought to shape according to the templets. They are grooved on the face to facilitate the even distribution of the abrasive, emery or fine carborundum, and the work is generally done on a special grinding machine, which moves the tool over the firmly supported disc in a complicated series of strokes imitating more or less closely the strokes found to be most effective in hand polishing.

In general terms the operator in handwork at this task supports the disc on a firm vertical post, by cementing it to a suitable holder, and then moves the tool over it in a series of straight or oval strokes, meanwhile walking around the post. A skilful operator watches the progress of his work, varies the length and position of his strokes accordingly, and, despite the unavoidable wear on the tool, can both keep its figure true and impart a true figure to the glass.

The polishing machine, of which a type used by Dr. Draper is shown in Fig. 44, produces a similar motion, the disc slowly revolving and the rather small tool moving over it in oval strokes kept off the center. More often the tool is of approximately the same diameter as the disc under it. The general character of the motion is evident from the construction. The disc *a* is chucked by *c c′* on the bed, turned by the post *d* and worm wheel *e*. This is operated from the pulleys, *i*, *g*, which drive through *k*. the crank *m*, adjustable in throw by the nuts *n*, *n′*, and in position of tool by the clamps *r*, *r*. The motion may be considerably

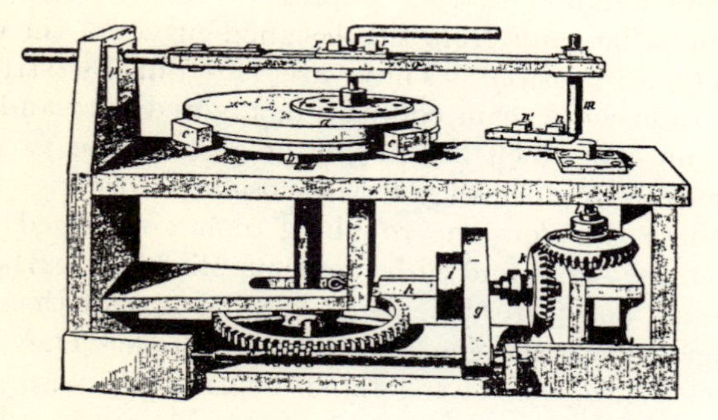

Fig. 44.—Dr. Draper's Polishing Machine.

varied by adjustment of the machine, always keeping the stroke from repeating on the same part of the disc, by making the period of the revolution and of the stroke incommensurable so far as may be. Even in spectacle grinding machines the stroke may repeat only once in hundreds of times, and even this frequency in a big objective would, if followed in the polishing, leave tool marks which could be detected in the final testing.

In the fine grinding, especially near the end of the process, the templets do not give sufficient precision in testing the curves, and recourse is had to the spherometer, by which measurements down to about $1/100000$ inch can be consistently made.

The next stage of operations is polishing, which transforms the grey translucency of the fine ground lens into the clear and brilliant surface which at last permits rigorous optical tests to be used for the final finish of the lens. This polishing is done generally on the fine grinding machine but with a very different tool and with rouge of the utmost fineness.

The polishing tool is in any case ground true and is then faced with a somewhat yielding material to carry the charge of rouge. Cheap lenses are commonly worked on a cloth polisher, a texture similar to billard cloth being suitable, or sometimes on paper worked dry.

With care either may produce a fairly good surface, with, however, a tendency to polish out the minute hollows left by grinding rather than to cut a true surface clear down to their bottoms. Hence cloth or paper is likely to leave microscopic inequalities apparently polished, and this may be sufficient to scatter over the field a very perceptible amount of light which should go to forming the image. All first class objectives and mirrors are in fact polished on optician's pitch. This is not the ordinary pitch of commerce but a substance of various composition, sometimes an asphaltic compound, again on a base of tar, or of resin brought to the right consistency by turpentine.

Whatever the exact composition, the fundamental property is that the material, apparently fairly hard and even brittle when cold, is actually somewhat plastic to continued pressure. Sealing wax has something of this quality, for a stick which may readily be broken will yet bend under its own weight if supported at the ends.

If the fine grinding process has been properly carried out the lens has received its correct form as nearly as gauges and the spherometer can determine it. The next step is to polish the surface as brilliantly and evenly as possible. To this end advantage is taken of the plastic quality already mentioned, that the glass may form its own tool.

The base of the tool may be anything convenient, metal, glass or even wood. Its working surface is made as nearly of the right curvature as practicable and it is then coated with warm pitch to a thickness of an eighth of an inch more or less, either continuously or in squares, and while still slightly warm the tool is placed against the fine ground disc, the exact shape of which it takes.

When cold the pitch surface can easily be cut out into squares or symmetrically pitted with a suitable tool, at once facilitating the distribution of the rouge and water that serves for polishing, and permitting delicate adjustment of the working curvature in a way about to be described.

Fig. 45 shows the squared surface of the tool as it would be used for polishing a plane or very slightly convex or concave surface. Supplied with the thin abrasive paste, it is allowed to settle, cold, into its final contact with the glass, and then the process of polishing by hand or machine is started.

The action of the tool must be uniform to avoid changing the shape of the lens. It can be regulated as it was in the grinding, by varying the length and character of the stroke, but even more delicately by varying the extent of surface covered by the pitch actually working on the glass.

This is done by channeling or boring away pitch near the rim or center of the tool as the case may be. Fig. 46 shows a

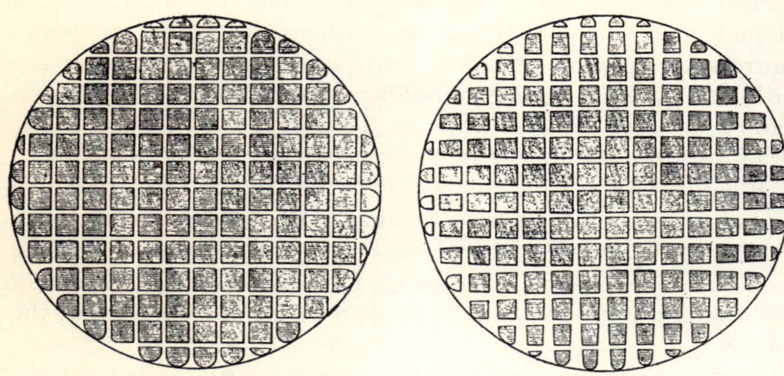

Fig. 45.—Tool for Flat Surface. Fig. 46.—Tool for Concave Surface.

tool which has been thus treated so that the squares are progressively smaller near the periphery. Such a spacing tends to produce a concave surface from a flat tool or to increase the concavity from a curved one. Trimming down the squares towards the centre produces the opposite result.

Broadly, the principle is that the tool cuts the more in the areas where the contact surfaces are the greater. This is not wholly by reason of greater abrading surface, but also because where the contact is greater in area the pitch settles less, from the diminished pressure, thus increasing the effective contact.

Clearly the effect of trimming away is correlated with the form and length of stroke, and the temper of the pitch, and in fact it requires the wisdom of the serpent to combine these various factors so as to produce the perfectly uniform and regular action required in polishing. Now and then, at brief intervals, the operation is stopped to supply rouge and to avoid changing the conditions by the heat of friction. Especially must heating be looked out for in hand polishing of lenses which is often done with the glass uppermost for easier inspection of the work.

Polishing, if the fine grinding has been judiciously done is, for moderate sized surfaces, a matter of only a few hours. It proceeds quite slowly at first while the hills are being ground down and then rather suddenly comes up brilliantly as the polisher reaches the bottoms of the valleys. Large lenses and mirrors may require many days.

Now begins the final and extraordinarily delicate process of figuring. The lens or mirror has its appointed form as nearly as the most precise mechanical methods can tell—say down to one or two hundred-thousandths of an inch. From the optical standpoint the result may be thoroughly bad, for an error of a few millionths of an inch may be serious in the final performance.

The periphery may be by such an amount longer or shorter in radius than it should be, or there may be an intermediate zone that has gone astray. In case of a mirror the original polishing is generally intended to leave a spherical surface which must be converted into a paraboloidal one by a change in curvature totalling only a few hundred-thousandths of an inch and seriously affected by much smaller variations.

The figuring is done in a fashion very similar to the polishing. The first step is to find out by optical tests such as are described in Chapter IX the location of the errors existing after the polishing, and once found, they must be eliminated by patient and cautious work on the surface.

Every optical expert has his own favorite methods of working out the figure. If there is a hollow zone the whole surface must be worked down to its level by repolishing; if, on the other hand, there is an annular hump, one may repolish with stroke and tool-face adapted to cut it down, or one may cautiously polish it out until it merges with the general level.

Polishing is commonly done with tools of approximately the size of the work, but in figuring there is great difference of practice, some expert workers depending entirely on manipulation of a full sized tool, others working locally with small polishers, even with the ball of the thumb, in removing slight aberrations. In small work where the glass can be depended on for homogeneity and the tools are easily kept true the former method is the usual one, but in big objectives the latter is often easier and may successfully reach faults otherwise very difficult to eliminate.

Among well known makers of telescopes the Clarks and their equally skilled successors the Lundins, father and son, developed the art of local retouching to a point little short of wizardry; the late Dr. Brashear depended almost entirely on the adroitly used polishing machine; Sir Howard Grubb uses local correction in certain cases, and in general the cautiously modified polisher; while some of the Continental experts are reported to have developed the local method very thoroughly.

The truth probably is that the particular error in hand should determine the method of attack and that its success depends entirely on the skill of the operator. As to the perfection of the objectives figured in either way, no systematic difference due to the method of figuring can be detected by the most delicate tests.

In any case the figuring operation is long and tedious, especially in large work where problems of supporting to avoid flexure arise, where temperature effects on tool and glass involve long delays between tests and correction, and where in the last resort non-spherical surfaces must often be resorted to in bringing the image to its final perfection.

The final test of goodness is performance, a clean round image without a trace of spherical or zonal aberration and the color correction the best the glasses will allow. Constant and rigorous testing must be applied all through the process of figuring, and the result seems to depend on a combination of experience, intuition and tactual expertness rarely united in any one person.

Sir Howard Grubb, in a paper to be commended to anyone interested in objectives, once forcibly said: "I may safely say that I have never finished any objective over 10 inches diameter, in the working of which I did not meet with some new experience, some new set of conditions which I had not met before, and which had then to be met by special and newly devised arrangements."

The making of reflecting telescopes is not much easier since although only one surface has to be worked, that one has to be figured with extraordinary care, flexure has to be guarded against at every stage of the working, and afterwards, temperature change is a busy foe, while testing for correct figure, the surface being non-spherical, is considerably more troublesome.

An expert can make a good mirror with far less actual labor than an objective of similar aperture, but when one reads Dr.

Henry Draper's statement that in spite of knowing at first hand the methods and grinding machines of Lord Rosse and Mr. Lassell, he ground over a hundred mirrors, and spent three years of time, before he could get a correct figure with reasonable facility, one certainly gains a high respect for the skill acquired.

This chapter is necessarily sketchy and not in the least intended to give the reader a complete account of technical glass manufacture, far less of the intricate and almost incommunicable art of making objectives and mirrors. It may however lead to a better understanding of the difference between the optical glass industry and the fabrication of commercial glass, and lead the reader to a fuller realization of how fine a work of art is a finished objective or mirror as compared with the crude efforts of the early makers or the hasty bungling of too many of their successors.

For further details on making, properties and working of optical glass see:

FOUCAULT: Annales de l'Obs. de Paris (Mémoires) V.5 (1859).
HOVESTADT: Jenaer Glas.
ROSENHAIN: Glass Manufacture.
SIR HOWARD GRUBB: Telescopic Objectives and Mirrors: Their Preparation and Testing. Nature **34**, 85.
DR. HENRY DRAPER: On the Construction of a Silvered Glass Telescope. (Smithsonian Contributions to Knowledge, Vol. 34.)
G. W. RITCHEY: On the Modern Reflecting Telescope and the Making and Testing of Optical Mirrors. (Smithsonian Contributions to Knowledge, Vol. 34.)
LORD RAYLEIGH: Polishing of Glass Surfaces. (Proc. Opt. Convention, 1905, p. 73.)
ORFORD: Lens Work for Amateurs.
REV. C. D. P. DAVIES: On the Testing of Parabolic Mirrors. M.N. **69**, 355.

CHAPTER IV

THE PROPERTIES OF OBJECTIVES AND MIRRORS

The path of the rays through an ordinary telescope has been shown in Fig. 5. In principle all the rays from a point in the distant object should unite precisely in a corresponding point in the image which is viewed by the eyepiece. Practically it takes very careful design and construction of the objective to make them meet in such orderly fashion even over an angular space of a single degree, and the wider the view required the more difficult the construction. We have spoken in the account of the early workers of their struggles to avoid chromatic and spherical aberrations, and it is chiefly these that still, in less measure, worry their successors.

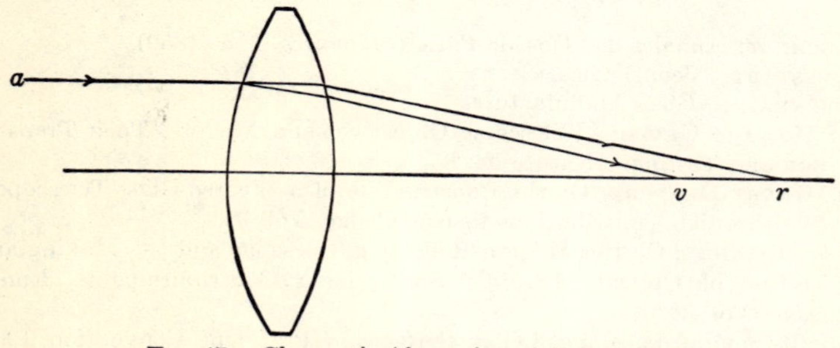

Fig. 47.—Chromatic Aberration of Convex Lens.

The first named is due to the fact that a prism does not bend light of all colors equally, but spreads them out into a spectrum; red refracted the least, violet the most. Since a lens may be regarded as an assemblage of prisms, of small angle near the centre and greater near the edge, it must on the whole and all over bend the blue and violet rays to meet on the axis nearer the rear surface than the corresponding red rays, as shown in Fig. 47. Here the incident ray a is split up by the prismatic effect of the lens, the red coming to a focus at r, the violet at v.

One can readily see this chromatic aberration by covering up most of a common reading glass with his hand and looking through the edge portion at a bright light, which will be spread out into a colored band.

If the lens is concave the violet rays will still be the more bent, but now outwards, as shown in Fig. 48. The incident ray a' is split up and the violet is bent toward v, proceeding as if coming straight from a virtual focus v' in front of the lens, and nearer it than the corresponding red focus r'. Evidently if we could combine a convex lens, bending the violet inward too much, with a concave one, bending it outward too much, the two opposite variations might compensate each other so that red and violet would come to the same focus—which is the principle of the achromatic objective.

If the refractive powers of the lenses were exactly proportional to their dispersive powers, as Newton erroneously thought, it is evident that the concave lens would pitch all the rays outwards

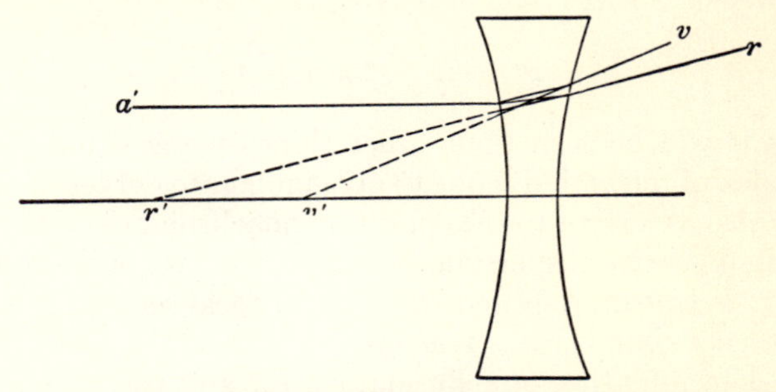

Fig. 48.—Chromatic Aberration of Concave Lens.

to an amount which would annul both the chromatic variation and the total refraction of the convex lens, leaving the pair without power to bring anything to a focus. Fortunately flint glass as compared with crown glass has nearly double the dispersion between red and violet, and only about 20% greater refractive power for the intermediate yellow ray.

Hence, the prismatic dispersive effect being proportional to the total curvature of the lens, the chromatic aberration of a crown glass lens will be cured by a concave flint lens of about half the total curvature, and, the refractions being about as 5 to 6, of $3/5$ the total power.

Since the "power" of any lens is the reciprocal of its focal length, a crown glass convex lens of focal length 3, and a concave flint lens of focal length 5 (negative) will form an approximately achromatic combination. The power of the combination will be the algebraic sum of the powers of the components so that the focal length of the pair will be about 5/2 that of the crown lens with which we started.

To be more precise the condition of achromatism is

$$\Sigma\rho\delta n + \Sigma\rho'\delta n' = 0$$

where ρ is the reciprocal of a radius and δn, or $\delta n'$, is the difference in refractive index between the rays chosen to be brought to exact focus together, as the red and the blue or violet.

This conventional equation simply states that the sum of the reciprocals of the radii of the crown lens multiplied by the dispersion of the crown, must equal the corresponding quantity for the flint lens if the two total dispersions are to annul each other, leaving the combination achromatic. Whatever glass is used the power of a lens made of it is

$$P\left(=\frac{1}{f}\right) = \Sigma\rho(n - 1)$$

so that it will be seen that, other things being equal, a glass of high index of refraction tends to give moderate curves in an objective. Also, referring to the condition of achromatism, the greater the difference in dispersion between the two glasses the less curvatures will be required for a given focal length, a condition advantageous for various reasons.

The determination of achromatism for any pair of glasses and focal length is greatly facilitated by employing the auxiliary quantity ν which is tabulated in all lists of optical glass as a short cut to a somewhat less manageable algebraic expression. Using this we can figure achromatism for unity focal length at once,

$$P = \frac{\nu}{\nu - \nu'}, \qquad P' = \frac{\nu'}{\nu - \nu'} \qquad \nu = \frac{n_D - 1}{\delta n}$$

being the powers of the leading and following lenses respectively. The combined lens will bring the rays of the two chosen colors, as red and blue, to focus at the same point on the axis. It does not necessarily give to the red and blue images of an object the same exact size. Failure in this respect is known as chromatic difference of magnification, but the fault is small and may generally be neglected in telescope objectives.

We have now seen how an objective may be made achromatic and of determinate focal length, but the solution is in terms of the sums of the respective curvatures of the crown and flint lenses, and gives no information about the radii of the individual surfaces. The relation between these is all-important in the final performance.

For in a convex lens with spherical surfaces the rays striking near the edge, of whatever color, are pitched inwards too much

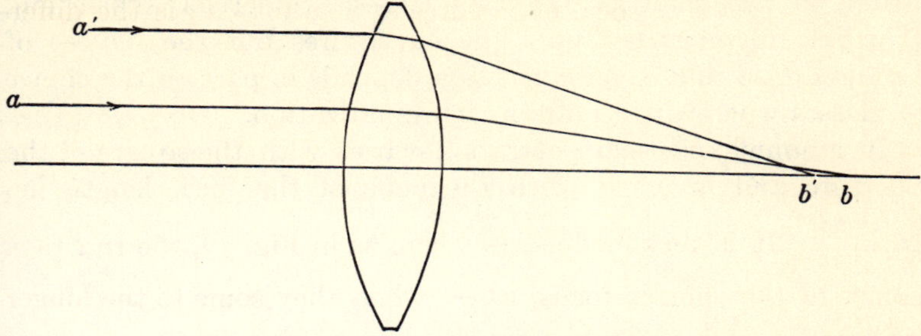

Fig. 49.—Spherical Aberration of Convex Lens.

compared with rays striking the more moderate curvatures near the axis, as shown in Fig. 49. The ray a' b' thus comes to a focus shorter than the ray a b.

This constitutes the fault of spherical aberration, which the old astronomers, following the suggestions of Descartes, tried ineffectually to cure by forming lenses with non-spherical surfaces.

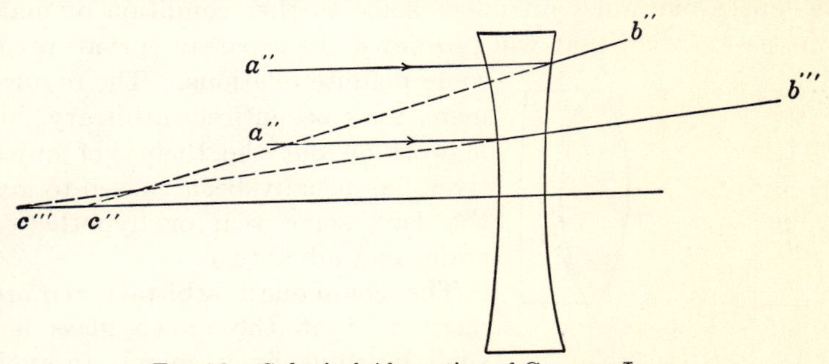

Fig. 50.—Spherical Aberration of Concave Lens.

Fig. 50 suggests the remedy, for the outer ray a'' is pitched out toward b'' as if it came from a focal point c'', while the ray nearer the center a''' is much less bent toward b''' as if it came from c'''. The spherical aberrations of a concave lens therefore, being opposite to those of a convex lens, the two must, at least to a certain extent, compensate each other as when combined in an achromatic objective.

So in fact they do, and, if the curves that go to make up the total curvatures of the two are properly chosen, the total spherical aberration can be made negligibly small, at least on and near the axis. Taking into account this condition, therefore, at once gives us a clue to the distribution of the total curvatures and hence to the radii of the two lenses. Spherical aberration, however, involves not only the curvatures but the indices of refraction, so that exact correction depends in part on the choice of glasses wherewith to obtain achromatization.

In amount spherical aberration varies with the square of the aperture and inversely with the cube of the focal length i.e. with $\frac{a^2}{f^3}$. It is reckoned as + when, as in Fig. 49, the rim rays come to the shorter focus, as −, when they come to the longer focus.

In any event, since the spherical aberration of a lens may be varied in above the ratio of 4:1, for the same total power, merely by changing the ratio of the radii, it is evident that the two lenses being fairly correct in total curvature might be given considerable variations in curvature and still mutually annul the axial spherical aberration.

Such is in fact the case, so that to get determinate forms for the lenses one must introduce some further condition or make some assumption that will pin down the separate curvatures to some definite relations. The requirement may be entirely arbitrary, but in working out the theory of objectives has usually been chosen to give the lens some real or hypothetical additional advantage.

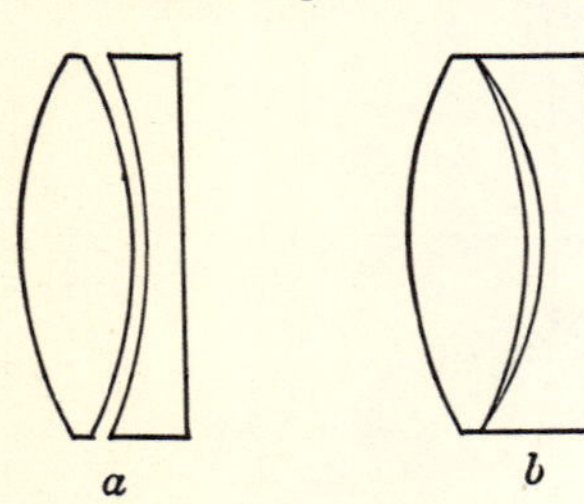

Fig. 51.—Objectives with Equiconvex Crown.

The commonest arbitrary requirement is that the crown glass lens shall be equiconvex, merely to avoid making an extra tool. This fixes one pair of radii, and the flint lens is then given the required compensating aberration choosing the easiest form to make. This results in the objective of Fig. 51.

Probably nine tenths of all objectives are of this general form, equiconvex crown and nearly or quite plano-concave flint. The inside radii may be the same, in which case the lenses should be cemented, or they may differ slightly in either direction as *a*, Fig. 51

with the front of the flint less curved than the rear of the crown, and *b* where the flint has the sharper curve. The resulting lens if ordinary glasses are chosen gives excellent correction of the spherical aberration on the axis, but not much away from it, yielding a rather narrow sharp field. Only a few exceptional combinations of glasses relieve this situation materially.

The identity of the inner radii so that the surfaces can be cemented is known historically as Clairault's condition, and since it fixes two curvatures at identity somewhat limits the choice of glasses, while to get proper corrections demands quite wide variations in the contact radii for comparatively small variations in the optical constants of the glass.

When two adjacent curves are identical they should be cemented, otherwise rays reflected from say the third surface of Fig. 51 will be reflected again from the second surface, and passing through the rear lens in almost the path of the original ray will come to nearly the same focus, producing a troublesome "ghost." Hence the curvatures of the second and third surfaces when not cemented are varied one way or the other by two or three per cent, enough to throw the twice reflected rays far out of focus.

In this case, as in most others, the analytical expression for the fundamental curvature to be determined turns up in the form of a quadratic equation, so that the result takes the form $a \pm \sqrt{b}$ and there are two sets of radii that meet the requirements. Of these the one presenting the gentler curves is ordinarily chosen. Fig. 52 *a* and *c* shows the two cemented forms, thus related, for a common pair of crown and flint glasses, both cleanly corrected for chromatic and axial spherical aberration.

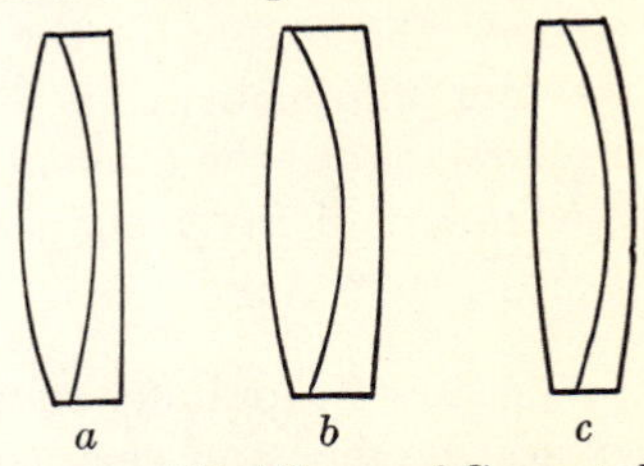

Fig. 52.—Allied Forms of Cemented Objectives.

Nearly a century ago Sir John Herschel proposed another defining condition, that the spherical aberration should be removed both for parallel incident rays and for those proceeding from a nearer point on the axis, say ten or more times the focal length in front of the objective. This condition had little practical value in itself, and its chief merit was that it approximated one that became of real importance if the second point were taken far enough away.

A little later Gauss suggested that the spherical aberration should be annulled for two different colors, much as the chromatic aberration is treated. And, being a mathematical wizard, he succeeded in working out the very intricate theory, which resulted in an objective approximately of the form shown in Fig. 53.

It does not give a wide field but is valuable for spectroscopic work, where keen definition in all colors is essential. Troublesome to compute, and difficult to mount and center, the type has not been much used, though there are fine examples of about 9½ inches aperture at Princeton, Utrecht, and Copenhagen, and a few smaller ones elsewhere, chiefly for spectroscopic use.

Fig. 53.—Gaussian Objective.

It was Fraunhofer who found and applied the determining condition of the highest practical value for most purposes. This condition was absence of *coma*, the comet shaped blur generally seen in the outer portions of a wide field.

It is due to the fact that parallel oblique rays passing through the opposite rims of the lens and through points near its center do not commonly come to the same focus, and it practically is akin to a spherical aberration for oblique rays which greatly reduces the extent of the sharp field. It is reckoned + when the blur points outwards, — when it points inwards, and is directly proportional to the tangent of the obliquity and the square of the aperture, and inversely to the square of the focal length i.e. it varies with $\frac{a^2 \tan u}{f^2}$.

Just how Fraunhofer solved the problem is quite unknown, but solve it he did, and very completely, as he indicates in one of his later papers in which he speaks of his objective as reducing all the aberrations to a minimum, and as Seidel proved 30 years later in the analysis of one of Fraunhofer's objectives. Very probably he worked by tracing axial and oblique rays through the objective form by trigonometrical computation, thus finding his way to a standard form for the glasses he used.[1]

[1] More recently his condition proves to be quite the exact equivalent of Abbé's *sine condition* which states that the sine of the angle made with the optical axis by a ray entering the objective from a given axial point shall bear a uniform ratio to the sine of the corresponding angle of emergence, whatever the point of incidence. For parallel rays along the axis this reduces to the requirement that the sines of the angles of emergence shall be proportional to the respective distances of the incident rays from the axis.

Fraunhofer's objective, of which Fig. 54*a* is an example worked by modern formulæ for the sine condition, gives very exact corrections over a field of 2°–3° when the glasses are suitably chosen and hence is invaluable for any work requiring a wide angle of view.

With certain combinations of glasses the coma-free condition may be combined successfully with Clairault's, although ordinarily the coma-free form falls between the two forms clear of spherical aberration, as in Fig. 52, *b*, which has its oblique rays well compensated but retains serious axial faults.

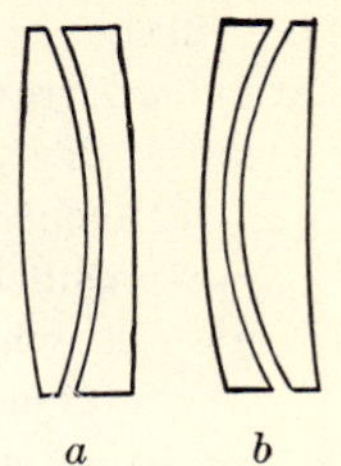

Fig. 54.—The Fraunhofer Types.

Fraunhofer's objective has for all advantageous combinations of glasses the front radius of the flint longer than the rear radius of the crown; hence the two must be separated by spacers at the edge, which in small lenses in simple cells is slightly inconvenient. However, the common attempt to simplify mounting by making the front flint radius the shorter almost invariably violates the sine condition and reduces the sharp field, fortunately not a very serious matter for most astronomical work.

The only material objection to the Fraunhofer type is the strong curvature of the rear radius of the crown which gives a form somewhat susceptible to flexure in large objectives. This is met in the flint-ahead form, developed essentially by Steinheil, and used in most of the objectives of his famous firm. Fig. 54*b* shows the flint-ahead objective corresponding to Fig. 54*a*. Obviously its curves are mechanically rather resistant to flexure.[1]

Mechanical considerations are not unimportant in large objectives, and Fig. 55, a highly useful form introduced by the Clarks and used in recent years for all their big lenses, is a case in point. Here there is an interval of about the proportion shown between the crown and flint components.

Fig. 55.—Clark Objective.

This secures effective ventilation allowing the lenses to come quickly to their steady temperature, and enables the inner

[1] It is interesting to note that in computing Fig. 54*a* for the sine condition, the other root of the quadratic gave roughly the Gaussian form of Fig. 53.

surfaces to be cleaned readily and freed of moisture. Optically it lessens the deviation from the sine condition otherwise practically inseparable from the equiconvex crown, reduces the chromatic difference of spherical aberration, and gives an easy way of controlling the color correction by slightly varying the separation of the lenses.

One further special case is worth noting, that of annulling the spherical aberration for rays passing through the lens in both directions. By proper choice of glass and curvatures this can be accomplished to a close approximation and the resulting form is shown in Fig. 56. The front of the crown is notably flat and the rear of the flint conspicuously curved, the shape in fact being intermediate between Figs. 52*b* and 52*c*. The type is useful in reading telescopes and the like, and for some spectroscopic applications.

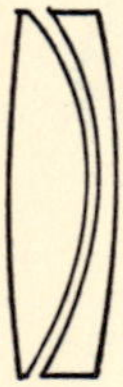

FIG. 56.—Corrected in Both Directions.

There are two well known forms of aberration not yet considered; astigmatism and curvature of field. The former is due to the fact that when the path of the rays is away from the axis, as from an extended object, those coming from a line radial to the axis, and those from a line tangent to a circle about the axis, do not come to the same focus. The net result is that the axial and tangential elements are brought to focus in two coaxial surfaces touching at the axis and departing more and more widely from each other as they depart from it. Both surfaces have considerable curvature, that for tangential lines being the sharper.

It is possible by suitable choice of glasses and their curvatures to bring both image surfaces together into an approximate plane for a moderate angular space about the axis without seriously damaging the corrections for chromatic and spherical aberration. To do this generally requires at least three lenses, and photographic objectives thus designed (*anastigmats*) may give a substantially flat field over a total angle of 50° to 60° with corrections perfect from the ordinary photographic standpoint.

If one demands the rigorous precision of corrections called for in astronomical work, the possible angle is very much reduced. Few astrographic lenses cover more than a 10° or 15° field, and the wider the relative aperture the harder it is to get an anastigmatically flat field free of material errors. Astigmatism is rarely noticeable in ordinary telescopes, but is sometimes conspicuous in eyepieces.

Curvature of field results from the tendency of oblique rays in objectives, otherwise well corrected, to come to shorter focus than axial rays, from their more considerable refraction resulting from greatly increased angles of incidence. This applies to both the astigmatic image surfaces, which are concave toward the objective in all ordinary cases.

Fortunately both of these faults are negligible near the axis. They are both proportional to $\frac{\tan^2 u}{f}$ where u is the obliquity to the axis and f the focal length; turn up with serious effect in wide

FIG. 57.—Steinheil Triple Objective. FIG. 58.—Tolles Quadruple Objective.

angled lenses such as are used in photography, but may generally be forgotten in telescopes of the ordinary *F* ratios, like *F*/12 to *F*/16. So also one may commonly forget a group of residual aberrations of higher orders, but below about *F*/8 look out for trouble. Objectives of wider aperture require a very careful choice of special glasses or the sub-division of the curvatures by the use of three or more lenses instead of two. Fig. 57 shows a cemented triplet of Steinheil's design, with a crown lens between two flints. Such triplets are made up to about 4 inches diameter and of relative apertures ranging from *F*/4 to *F*/5.

In cases of demand for extreme relative aperture, objectives composed of four cemented elements have now and then been produced. An example is shown in Fig. 58, a four-part objective of 1 inch aperture made by Tolles years ago for a small hand telescope. Its performance, although it worked at *F*/4, was reported to be excellent even up to 75 diameters.

The main difficulty with these objectives of high aperture is the relatively great curvature of field due to short focal length which prevents full utilization of the improved corrections off the axis.

Distortion is similarly due to the fact that magnification is not quite the same for rays passing at different distances from the axis. It varies in general with the cube of the distance from the axis, and is usually negligible save in photographic telescopes, ordinary visual fields being too small to show it conspicuously.

Distortion is most readily avoided by adopting the form of a symmetrical doublet of at least four lenses as in common photographic use. No simple achromatic pair gives a field wholly free of distortion and also of the ordinary aberrations, except very near the axis, and in measuring plates taken with such simple objectives corrections for distortion are generally required.

At times it becomes necessary to depart somewhat from the objective form which theoretically gives the least aberrations in order to meet some specific requirement. Luckily one may modify the ratios of the curves very perceptibly without serious results. The aberrations produced come on gradually and not by jumps.

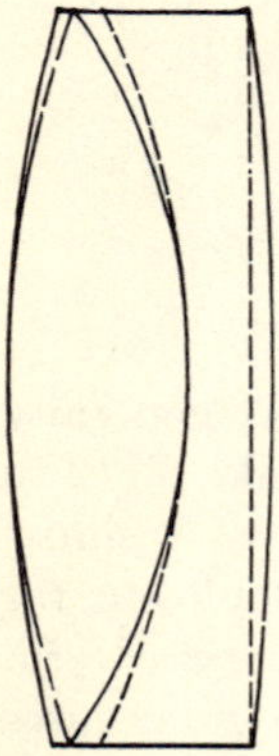

A case in point is that of the so-called "bent" objective in which the curvatures are all changed symmetrically, as if one had put his fingers on the periphery and his thumbs on the centre of the whole affair, and had sprung it noticeably one way or the other.

FIG. 59—"Bent" Objective.

The corrections in general are slightly deteriorated but the field may be in effect materially flattened and improved. An extreme case is the photographic landscape lens. Figure 59 is an actual example from a telescope where low power and very large angular view were required. The objective was first designed from carefully chosen glass to meet accurately the sine condition. Even so the field, which covered an apparent angle of fully 40°, fell off seriously at the edge.

Bearing in mind the rest of the system, the objective was then "bent" into the form given by the dotted lines, and the telescope then showed beautiful definition clear to the periphery of the field, without any visible loss in the centre.

This spurious flattening cannot be pushed far without getting into trouble for it does not cure the astigmatic difference of focus, but it is sometimes very useful. Practically curvature of field is an outstanding error that cannot be remedied in objectives required to stand high magnifying powers, except by going to the anastigmatic forms similar to those used in photography.[1]

Aside from curvature the chief residual error in objectives is imperfection of achromatism. This arises from the fact that crown and flint glasses do not disperse the various colors quite in the same ratio. The crown gives slightly disproportionate importance to the red end of the spectrum, the flint to the violet end—the so-called "irrationality of dispersion."

Hence if a pair of lenses match up accurately for two chosen colors like those represented by the C and F lines, they will fail of mutual compensation elsewhere. Figure 60 shows the situation. Here the spectra from crown and flint glasses are brought to exactly the same extent between the C and F lines, which serve as landmarks.

Clearly if two prisms or lenses are thus adjusted to the same refractions for C and F, the light passing through the combination will still be slightly colored in virtue of the differences elsewhere in the spectrum. These residual color differences produce what is known as the "secondary spectrum."

What this does in the case of an achromatic lens may be clearly seen from the figure; C and F having exactly the same refractions in the flint and crown, come to the same focus. For D, the yellow line of sodium, the flint lens refracts a shade the less, hence is not quite powerful enough to balance the crown, which therefore brings D to a focus a little shorter than C and F. On the other hand for A′ and G′, the flint refracts a bit more than the crown, overbalances it and brings these red and violet rays to a focus a little longer than the joint C and F focus.

[1] The curvature of the image is the thing which sets a limit to shortening the relative focus, as already noted, for the astigmatic image surfaces as we have seen, fall rapidly apart away from the axis, and both curvatures are considerable. The tangential is the greater, corresponding roughly to a radius notably less than 1/3 the focal length, while the radial fits a radius of less than 2/3 this length with all ordinary glasses, given forms correcting the ordinary aberrations. The curves are concave towards the objective except in "anastigmats" and some objectives having bad aberrations otherwise. Their approximate curvatures assuming a semiangular aperture for an achromatic objective not over say 5°, have been shown to be, to focus unity

$$\rho_r = 1 + \frac{1}{\nu - \nu'}\left(\frac{\nu}{n} - \frac{\nu'}{n'}\right), \text{ and } \rho_t = 3 + \frac{1}{\nu - \nu'}\left(\frac{\nu}{n} - \frac{\nu'}{n'}\right)$$

ρ_r and ρ_t being the respective reciprocals of the radii. The surfaces are really somewhat egg shaped rather than spherical as one departs from the axis.

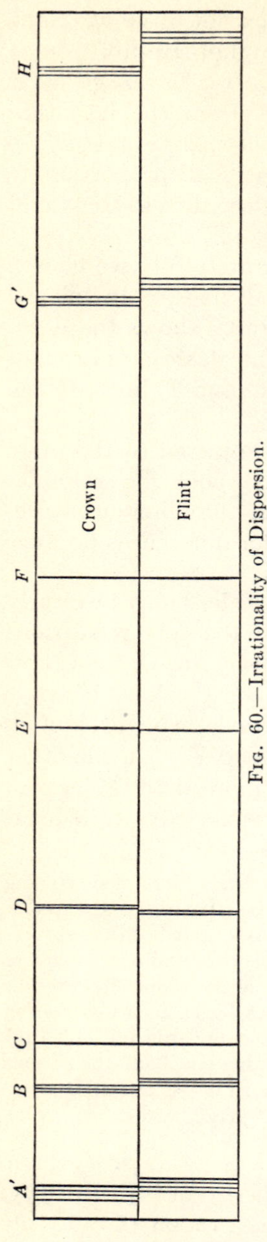

FIG. 60.—Irrationality of Dispersion.

The difference for D is quite small, roughly about 1/2000 of the focal length, while the red runs long by nearly three times that amount, the violet by about four. Towards the H line the difference increases rapidly and in large telescopes the actual range of focus for the various colors amounts to several inches.

This difficulty cannot be avoided by any choice among ordinary pairs of glasses, which are nearly alike in the matter of secondary spectrum. In the latter part of the last century determined efforts were made to produce glasses that would give more nearly an equal run of dispersion, at first by English experimenters, and then with final success by Schott and Abbé at Jena.

Both crown and flint had to be quite abnormal in composition, especially the latter, and the pair were of very nearly the same refractive index and with small difference in the quantity ν which we have seen determines the general amount of curvature. Moreover it proved to be extremely hard to get the crown quite homogeneous and it is listed by Schott with the reservation that it is not free from bubbles and striæ.

Nevertheless the new glasses reduce the secondary spectrum greatly, to about 1/4 of its ordinary value, in the average. It is difficult to get rid of the spherical aberration, however, from the sharp curves required and the small difference between the glasses, and it seems to be impracticable on this account to go to greater aperture than about $F/20$.

Figure 61 shows the deeply curved form necessary even at half the relative aperture usable with common glasses. At $F/20$ the secondary spectrum from the latter is not conspicuous and Roe (Pop. Ast. *18*, 193), testing side by side a small Steinheil of the new glasses, and a Clark of the old, of almost identical size and focal ratio, found no difference in their practical performance.

Another attack on the same problem was more successfully made by H. D. Taylor. Realizing the difficulty found with a doublet objective of even the best matched of the new glasses, he adopted the plan of getting a flint of exactly the right dispersion

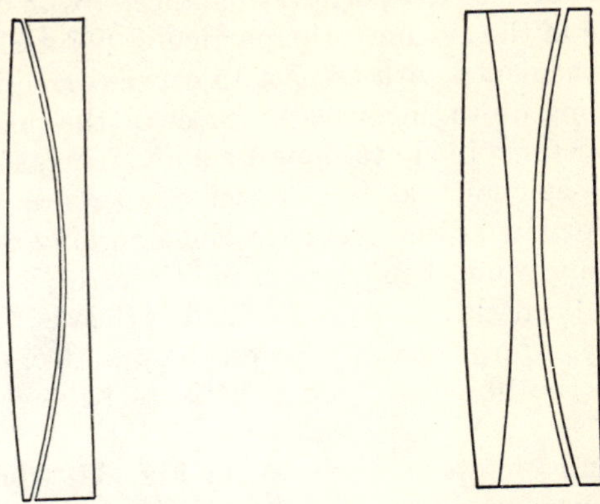

Fig. 61.—Apochromatic Doublet. Fig. 62.—Apochromatic Triplet.

by averaging the dispersions of a properly selected pair of flints formed into lenses of the appropriate relative curvatures.

The resulting form of objective is made, especially, by Cooke of York, and also by Continental makers, and carries the name of "photo-visual" since the exactness of corrections is carried well into the violet, so that one can see and photograph at the same focus. The residual chromatic error is very small, not above ⅛ to $\frac{1}{10}$ the ordinary secondary spectrum.

By this construction it is practicable to increase the aperture to $F/12$ or $F/10$ while still retaining moderate curvatures and reducing the residual spherical aberration. There are a round dozen triplet forms possible, of which the best, adopted by Taylor, is shown in Fig. 62. It has the duplex flint ahead—first a baryta light flint, then a borosilicate flint, and to the rear a special light crown The two latter glasses have been under some suspicion as to permanence, but the difficulty has of late years been reported as remedied. Be that as it may, neither doublets nor triplets with reduced secondary spectrum have come into any large use for astronomical purposes. Their increased cost is considerable,[1] their aperture even in the triplet, rather small for astrophotography, and their achromatism is still lacking the perfection reached by a mirror.

The matter of achromatism is further complicated by the fact that objectives are usually over-achromatized to compensate for the chromatic errors in the eyepiece, and especially in the eye. As a general rule an outstanding error in any part of an optical system can be more or less perfectly balanced by an opposite error anywhere else in the system—the particular point chosen being a matter of convenience with respect to other corrections.

The eye being quite uncorrected for color the image produced even by a reflector is likely to show a colored fringe if at all bright, the more conspicuous as the relative aperture of the pupil increases. For low power eyepieces the emerging ray may quite fill a wide pupil and then the chromatic error is troublesome. Hence it has been the custom of skilled opticians, from the time of Fraunhofer, who probably started the practice, to overdo the correction of the objective just a little to balance the fault of the eye.

What actually happens is shown in Fig. 63, which gives the results of achromatization as practiced by some of the world's adepts. The shortest focus is in the yellow green, not far from the line D. The longest is in the violet, and F, instead of coinciding in focus with C as it is conventionally supposed to do, actually coincides with the deep and faint red near the line marked B. Hence the visible effect is to lengthen the focus for blue enough to make up for the tendency of the eye in the other direction. The resulting image then should show no conspicuous rim of red or blue. The actual adjustment of the color correction is almost wholly a matter of skilled judgment but Fig. 63 shows that of the great makers to be quite uniform. The smallest overcorrection is found in the Grubb objective, the largest in the Fraunhofer. The differences seem to be due mainly to individual variations of opinion as to what diameter of pupil should be taken as typical for the eye.

The common practice is to get the best possible adjustment for a fairly high power, corresponding to a beam hardly $\frac{1}{64}$ inch in diameter entering the pupil.

[1] The doublet costs about one and a half times, and the triplet more than twice the price of an ordinary achromatic of the same aperture.

In any case the bigger the pencil of rays utilized by the eye, i.e., the lower the power, the more overcorrection must be provided, so that telescopes intended, like comet seekers, for regular

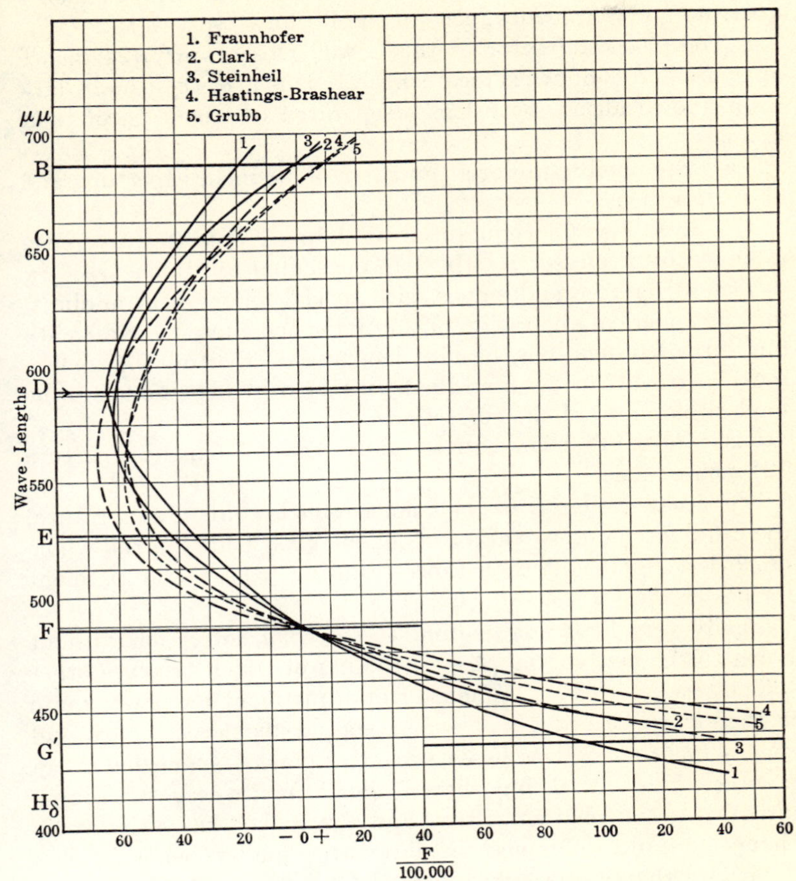

Fig. 63.—Achromatization Curves by Various Makers.

use with low powers must be designed accordingly, either as respects objective or ocular.

The differences concerned in this chromatic correction for power are by no means negligible in observing, and an objective actually conforming to the C to F correction assumed in tables of optical glass would produce a decidedly unpleasant impression when used with various powers on bright objects. And the values for ν implied in the actual color correction are not immaterial in computing the best form for a proposed objective.

1 is from Franunhofer's own hands, the instrument of 9.6 inches aperture and 170 inches focus in the Berlin Observatory.

2 The Clark refractor of the Lowell Observatory, 24 inches aperture and 386 inches focal length. This is of the usual Clark form, crown ahead, with lenses separated by about 1/6 of their diameter.

3 is a Steinheil refractor at Potsdam of 5.3 inches aperture, and 85 inches focus.

4 is from the fine equatorial at Johns Hopkins University, designed by Professor Hastings and executed by Brashear.

The objective was designed with special reference to minimizing the spherical aberration not only for one chosen wave length but for all others, has the flint lens ahead, aperture 9.4 inches, focal length 142 inches, and the lenses separated by 1/4 inch in the final adjustment of the corrections.

5 is from the Potsdam equatorial by Grubb, 8.5 inches aperture 124 inches focus.

The great similarity of the color curves is evident at a glance, the differences due to variations in the glass being on the whole much less significant than those resulting from the adjustment for power.

Really very little can be done to the color correction without going to the new special glasses, the use of which involves other difficulties, and leaves the matter of adjustment for power quite in the air, to be brought down by special eyepieces. Now and then a melting of glass has a run of dispersion somewhat more favorable than usual, but there is small chance of getting large discs of special characteristics, and the maker has to take his chance, minute differences in chromatic quality being far less important than uniformity and good annealing.

Regarding the aberrations of mirrors something has been said in Chap. I, but it may be well here to show the practical side of the matter by a few simple illustrations.

Figure 64 shows the simplest form of concave mirror—a spherical surface, in this instance of 90° aperture, the better to show its properties. If light proceeded radially outward from C, the center of curvature of the surface, evidently any ray would strike the surface perpendicularly as at a and would be turned squarely back upon itself, passing again through the center of curvature as indicated in the figure.

A ray, however, proceeding parallel to the axis and striking the surface as at *bb* will be deflected by twice the angle of incidence as is the case with all reflected rays. But this angle is measured by the radius *Cb* from the center of curvature and the reflected ray makes an angle *CbF* with the radius, equal to *FCb*.

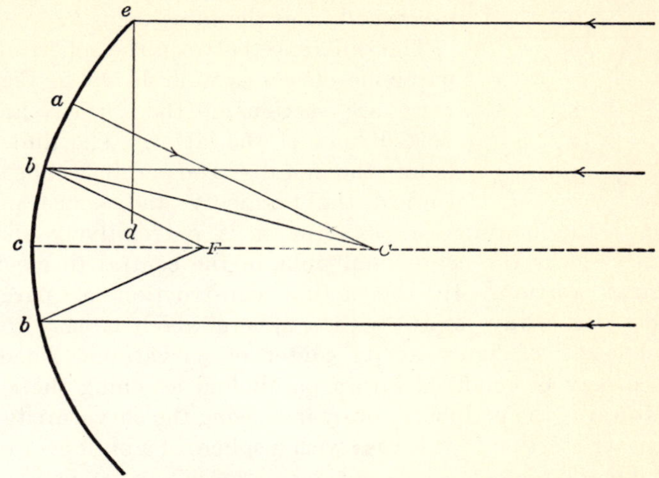

Fig. 64.—Reflection from Concave Spherical Mirror.

For points very near the axis *bF*, therefore, equals *FC*, and substantially also equals *cF*. Thus rays near the axis and parallel to it meet at *F* the focus half, way from *c* to *C*. The equivalent focal length of a spherical concave mirror of small aperture is therefore half its radius of curvature.

But obviously for large angles of incidence these convenient equalities do not hold. As the upper half of the figure shows, the ray parallel to the axis and incident on the mirror 45° away at *e* is turned straight down, for it falls upon a surface inclined to it by 45°and is therefore deflected by 90°, cutting the axis far inside the nominal focus, at *d*. Following up other rays nearer the axis it appears that there is no longer a focal point but a cusp-like focal surface, known to geometrical optics as a caustic and permitting no well defined image.

A paraboloidal reflecting surface as in Fig. 65 has the property of bringing to a single point focus all rays parallel to its axis while quite failing of uniting rays proceeding from any point on its axis, since its curvature is changing all the way out from vertex to periphery. Here the parallel rays a, a, a, a meeting the surface are reflected to the focus F, while in a perfectly symmetrical way the prolongation of these rays a', a', a', a' if incident on the convex surface of the paraboloid would be reflected in R, R', R'', R''' just as if they proceeded from the same focus F.

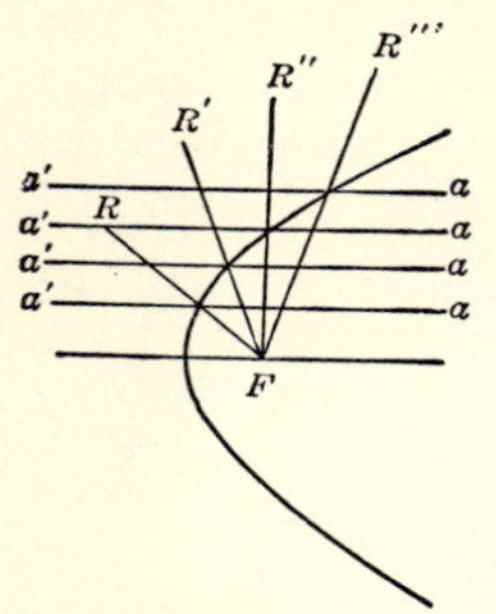

Fig. 65.—Reflection from Paraboloid.

The difference between the spherical and parabolic curves is well shown in Fig. 66. Here are sections of the former, and in dotted lines of the latter. The difference points the moral. The parabola falls away toward the periphery and hence pushes outward the marginal rays. But it is of relatively sharper curvature near the center and pulls in the central to meet the marginal portion. In the actual construction of parabolic mirrors one always starts with a sphere which is easy to test for precision of figure at its center of curvature. Then the surface may be modified into a paraboloid lessening the curvature towards the periphery, or by increasing the curvature toward the center starting in this case with a sphere of a bit longer radius as in Fig. 66a.

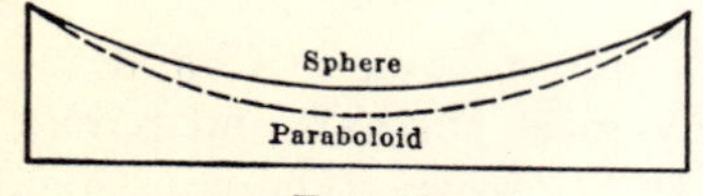

Fig. 66a.

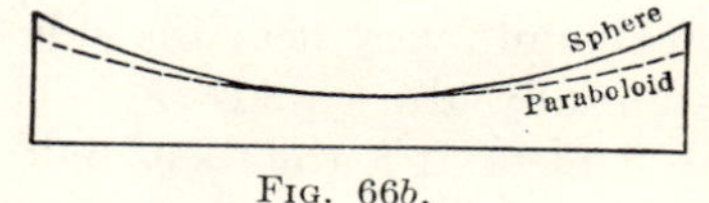

Fig. 66b.

Variation of Paraboloid from Sphere.

Practice differs in this respect, either process leading to the same result. In any case the departure from the spherical curve is very slight—a few hundred thousandths or at most ten thousandths of an inch depending on the size and relative focus of the mirror.

Yet this small variation makes all the difference between admirable and hopelessly bad definition. However the work is done it is guided by frequent testing, until the performance shows that a truly parabolic figure has been reached. Its attainment is a matter of skilled judgment and experience.

The weak point of the parabolic mirror is in dealing with rays coming in parallel but oblique to the axis. Figure 67 shows the situation plainly enough. The reflected rays a', a'' no longer meet in a point at the focus F but inside the focus for parallel rays, at f forming a surface of aberration. The practical effect is that the image rapidly deteriorates as the star passes away from the axis, taking on an oval character that suggests a bad case of astigmatism with serious complications from coma, which in fact is substantially the case.

Even when the angular aperture is very small the focal surface is nevertheless a sphere of radius equal to twice the focal length,

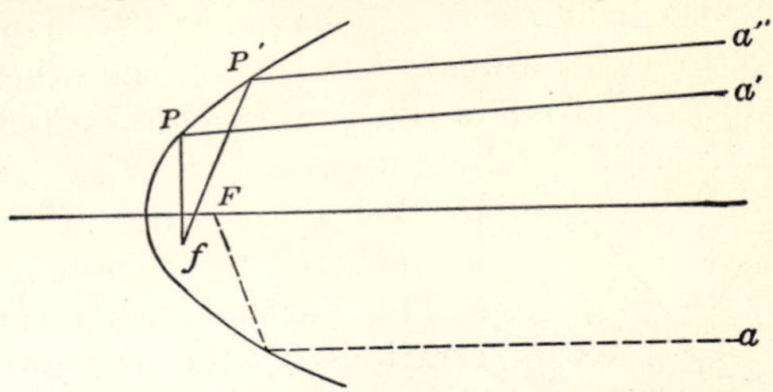

Fig. 67.—Aberration of Parabolic Mirror.

and the aberrations off the axis increase approximately as the square of the relative aperture, and directly as the angular distance from the axis.

The even tolerably sharp field of the mirror is therefore generally small, rarely over 30′ of arc as mirrors are customarily proportioned. At the relative aperture usual with refractors, say F/15, the sharp fields of the two are quite comparable in extent.

The most effective help for the usual aberrations[1] of the

[1] A very useful treatment of the aberrations of parabolic mirrors by Poor is in Ap. J. **7**, 114. In this is given a table of the maximum dimension of a star disc off the axis in reflectors of various apertures. This table condenses to the closely approximate formula

$$a = \frac{11d}{f^2}$$

where a is the aberrational diameter of the star disc, in seconds of arc, d the distance from the axis in minutes of arc, f the denominator of the F ratio (F/8 &c) and 11, a constant. Obviously the separating power of a telescope (see Chap. X) being substantially $\frac{4.''56}{D}$ where D is the diameter of objective or mirror in inches, the separating power will be impaired when $a > \frac{4.''56}{D}$. In the photographic case the critical quantity is not $\frac{4.''56}{D}$, but the maximum image diameter tolerable for the purpose in hand.

mirror is the adoption of the Cassegrain form, by all odds the most convenient for large instruments, with a hyperboloidal secondary mirror.

The hyperboloid is a curve of very interesting optical properties. Just as a spherical mirror returns again rays proceeding from its center of curvature without aberration, and the paraboloid sends from its focus a parallel axial beam free of aberration, or returns such a beam to an exact focus again, so a hyperboloid, Fig. 68, sends out a divergent beam free from aberration or brings it, returning, to an exact focus.

Such a beam *a*, *a*, *a*, in fact behaves as if it came from and returned to a virtual conjugate focus F' on the other side of the hyperbolic surface. And if the convex side be reflecting, converging rays R, R', R'', falling upon it at P, P', P'', as if headed for the virtual focus F, will actually be reflected to F', now a real focus.

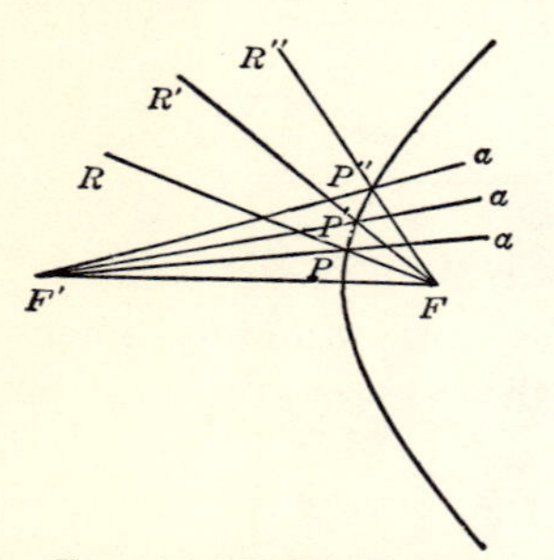

Fig. 68.—Reflection from Hyperboloid.

This surface being convex its aberrations off the axis are of opposite sign to those due to a concave surface, and can in part at least, be made to compensate the aberrations of a parabolic main mirror. The rationale of the operation appears from comparison of Figs. 67 and 68.

In the former the oblique rays *a*, a' are pitched too sharply down. When reflected from the convex surface of Fig. 68 as a converging beam along R, R', R'', they can nevertheless, if the hyperbola be properly proportioned, be brought down to focus at F' conjugate to F, their approximate mutual point of convergence.

Evidently, however, this compensation cannot be complete over a wide angle, when F' spreads into a surface, and the net result is that while the total aberrations are materially reduced there is some residual coma together with some increase of curvature of field, and distortion. Here just as in the parabolizing of the large speculum the construction is substantially empirical, guided in the case of a skilled operator by a sort of insight derived from experience.

Starting from a substantially spherical convexity of very nearly the required curvature the figure is gradually modified as in the earlier example until test with the truly parabolic mirror shows a flawless image for the combination. The truth is that no conic surface of revolution save the sphere can be ground to true figure by any rigorous geometrical method. The result must depend on the skill with which one by machine or hand can gauge minute departures from the sphere.

Attempts have been made by the late Professor Schwarzchild and others to improve the corrections of reflectors so as to increase the field but they demand either very difficult curvatures imposed on both mirrors, or the interposition of lenses, and have thus far reached no practical result.

References

SCHWARZCHILD: Untersuchungen Geom., Opt. II.
SAMPSON: *Observatory 36*, 248.
CODDINGTON: Reflexion and Refraction of Light.
HERSCHEL: Light.
TAYLOR: Applied Optics.
SOUTHALL: Geometrical Optics.
MARTIN: *Ann. Sci. de l'Ecole Normale*, 1877, Supplement.
MOSER: *Zeit. f. Instrumentenkunde*, 1887.
HARTING: *Zeit. f. Inst.*, 1899.
HARTING: *Zeit. f. Inst.*, 1898.
VON HOEGH: *Zeit. f. Inst.*, 1899.
STEINHEIL & VOIT: Applied Optics.
COLLECTED RESEARCHES, National Physical Laboratory, Vol. 14.
GLEICHEN: Lehrbuch d. Geometrische Optik.

NOTE.—In dealing with optical formulæ look out for the algebraic signs. Writers vary in their conventions regarding them and it sometimes is as difficult to learn how to apply a formula as to derive it from the start. Also, especially in optical patents, look out for camouflage, as omitting to specify an optical constant, giving examples involving glasses not produced by any manufacturer, and even specifying curves leading to absurd properties. It is a good idea to check up the achromatization and focal length before getting too trustful of a numerical design.

CHAPTER V

MOUNTINGS

A steady and convenient mounting is just as necessary to the successful use of the telescope as is a good objective. No satisfactory observations for any purpose can be made with a telescope unsteadily mounted and not provided with adjustments enabling it to be moved smoothly and easily in following a celestial object.

Broadly, telescope mounts may be divided into two general classes, altazimuth and equatorial. The former class is, as its name suggests, arranged to be turned in azimuth about a vertical axis, and in altitude about a horizontal axis. Such a mounting may be made of extreme simplicity, but obviously it requires two motions in order to follow up any object in the field, for the apparent motion of the heavenly bodies is in circles about the celestial pole as an axis, and consequently inclined from the vertical by the latitude of the place of observation.

Pointing a telescope with motions about a vertical and horizontal axis only, therefore means that, as a star moves in its apparent path, it will drift away from the telescope both in azimuth and in altitude, and require to be followed by a double motion.

Altazimuth mounts may be divided into three general groups according to their construction. The first and simplest of them is the pillar-and-claw stand shown in Figure 69. This consists of a vertical pillar supported on a strong tripod, usually of brass or iron, and provided at its top with a long conical bearing carrying at its upper extremity a hinged joint, bearing a bar to which the telescope is screwed as shown in the illustration.

If properly made the upper joint comprises a circular plate carrying the bar and held between two cheek pieces with means for taking up wear, and providing just enough friction to permit of easy adjustment of the telescope, which can be swung in altitude from near the zenith to the horizon or below, and turned around its vertical axis in any direction.

When well made a stand of this kind is steady and smooth working, readily capable of carrying a telescope up to about 3 inches aperture. It needs for its proper use a firm sub-support for the three strong hinged legs of the pillar. This is conveniently

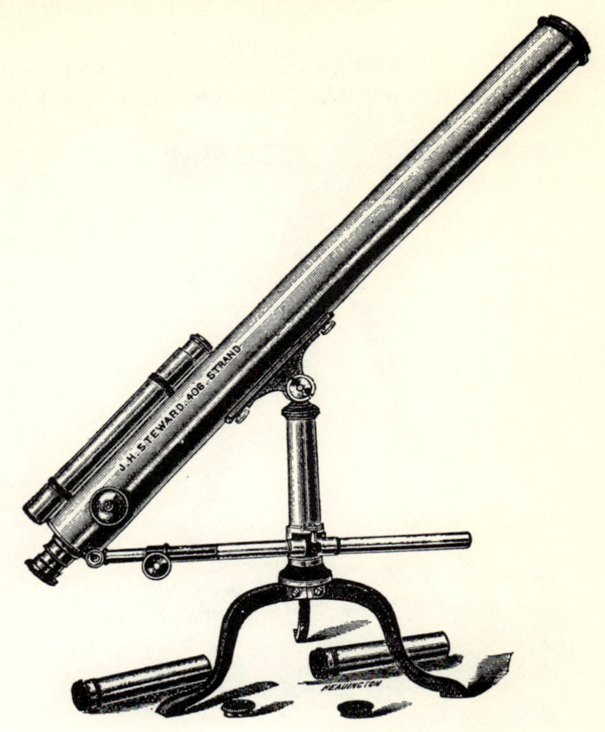

Fig. 69.—Table Mount with Slow Motion.

made as a very solid stool with spreading legs, or a plank of sufficient size may be firmly bolted to a well set post.

Thus arranged the mount is a very serviceable one for small instruments. Its stability, however, depends on the base upon which it is set. The writer once unwisely attempted to gain convenience by removing the legs of the stand and screwing its body firmly upon a very substantial tripod. The result was a complete failure in steadiness, owing to the rather long lever arm furnished by the height of the pillar; and the instrument, which had been admirably steady originally, vibrated abominably when touched for focussing.

The particular stand here shown is furnished with a rack motion in altitude which is a considerable convenience in following. More rarely adjustable steadying rods attached to the objective end of the instrument are brought down to its base, but for a telescope large enough to require this a better mount is generally desirable.

Now and then an altazimuth head of just the sort used in the pillar-and-claw stand is actually fitted on a tall tripod, but such

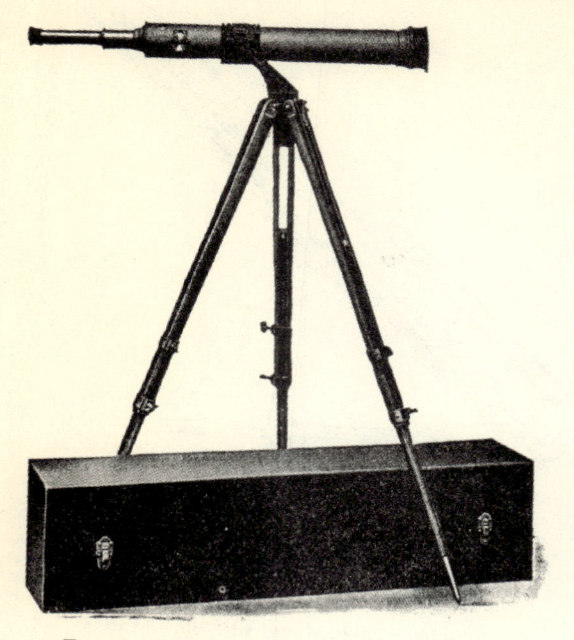

FIG. 70.—Altazimuth Mount, Clark Type T.

an arrangement is usually found only in cheap instruments and for such tripod mountings other fittings are desirable.

The second form of altazimuth mount, is altogether of more substantial construction. The vertical axis, usually tapered and carefully ground in its bearings, carries an oblique fork in which the telescope tube is carried on trunnions for its vertical motion. The inclination of the forked head enables the telescope to be pointed directly toward the zenith and the whole mount forms the head of a well made tripod.

Figure 70 shows an excellent type of this form of mount as

used for the Clark Type T telescope, designed for both terrestrial and astronomical observation. In this particular arrangement the telescope lies in an aluminum cradle carried on the trunnions,

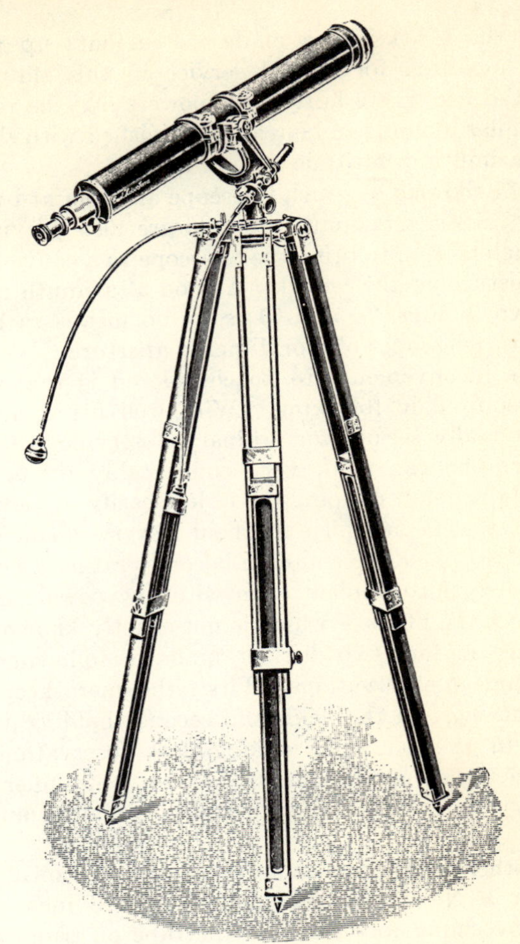

FIG. 71.—Altazimuth with Full Slow Motions.

from which it can be readily removed by loosening the thumb screws and opening the cradle.

It can also be set longitudinally for balance in the cradle if any attachments are to be placed upon either end. Here the adjustment for the height of the instrument is provided both in the spread of the tripod and in the adjustable legs. So mounted a telescope of 3 or 4 inches aperture is easily handled and capable of rendering very good service either for terrestrial or celestial work.

Indeed the Clarks have made instruments up to 6 inches aperture, mounted for special service in this simple manner. For celestial use where fairly high powers may be required this or any similar mount can be readily furnished with slow motions either in azimuth or altitude or both.

Figure 71 shows a 4¼-inch telescope and mount by Zeiss thus equipped. Some altazimuth mounts are also provided with a vertical rack motion to bring the telescope to a convenient height without disturbing the tripod. A good altazimuth mount such as is shown in Figs. 70 and 71 is by no means to be despised for use with telescopes of 3 or 4 inches aperture.

The sole inconvenience to be considered is that of the two motions required in following. With well fitted slow motions this is not really serious for ordinary observing with moderate powers, for one can work very comfortably up to powers of 150 or 200 diameters keeping the object easily in view; but with the higher powers in which the field is very small, only a few minutes of arc, the double motion becomes rather a nuisance and it is extremely inconvenient even with low powers in sweeping for an object the place of which is not exactly known.

There are in fact two distinct kinds of following necessary in astronomical observations. First, the mere keeping of the object somewhere in the field, and second, holding it somewhat rigorously in position, as in making close observations of detail or micrometrical measurements. When this finer following is necessary the sooner one gets away from altazimuth mounts the better.

Still another form of altazimuth mount is shown in Fig. 72 applied for a Newtonian reflector of 6 or 8 inches aperture. Here the overhung fork carrying the tube on trunnions is supported on a stout fixed tripod, to which it is pivoted at the front, and it is provided at the rear with a firm bearing on a sector borne by the tripod.

At the front a rod with sliding coarse, and screw fine, adjustment provides the necessary motion in altitude. The whole fork is swung about its pivot over the sector bearing by a cross screw turned by a rod with a universal joint.

This mount strongly suggests the original one of Hadley, Fig. 16, and is most firm and serviceable. A reflector thus mounted is remarkably convenient in that the eyepiece is always in a most accessible position, the view always horizontal, and the adjustments always within easy reach of the observer.

Fig. 72.—Altazimuth Newtonian Reflector.

Whenever it is necessary to follow an object closely, as in using a micrometer and some other auxiliaries, the altazimuth mount is troublesome and some modification adjustable by a single motion, preferably made automatic by clockwork, becomes necessary.

The first step in this direction is a very simple one indeed.

Suppose one were to tilt the azimuth axis so that it pointed to the celestial pole, about which all the stars appear to revolve. Then evidently the telescope being once pointed, a star could be followed merely by turning the tube about this tilted axis. Of course one could not easily reach some objects near the pole without, perhaps, fouling the mount, but in general the sky is within reach and a single motion follows the star, very easily if the original mount had a slow motion in azimuth.

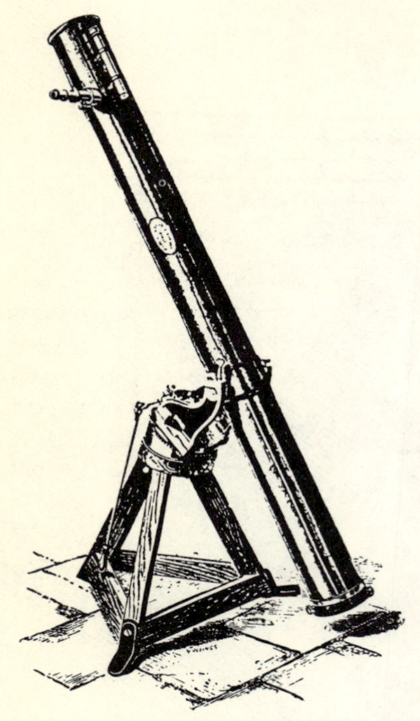

FIG. 73.—Parallactic Mount for Reflector.

This is in fact the simplest form of equatorial mount, sometimes called parallactic. Figure 73 shows the principle applied to a small reflector. An oblique block with its angle adjusted to the co-latitude of the place drops the vertical axis into line with the pole, and the major part of the celestial vault is then within easy reach.

It may be regarded as the transition step from the alt-azimuth to the true equatorial. It is rarely used for refractors, and the first attempt at a real equatorial mount was in fact made by James Short F. R. S. in mounting some of his small Gregorians.[1] As a matter of record this is shown, from Short's own paper before the Royal Society in 1749, in Fig. 74.

A glance shows a stand apparently most complicated, but closer examination discloses that it is merely an equatorial on a table stand with a sweep in declination over a very wide arc,

[1] Instruments with a polar axis were used by Scheiner as early as 1627; by Roemer about three quarters of a century later, and previously had been employed, using sights rather than telescopes, by the Chinese; but these were far from being equatorials in the modern sense.

and quite complete arrangements for setting to the exact latitude and azimuth. The particular instrument shown was of 4 inches aperture and about 18 inches long and was one of several produced by Short at about this epoch.

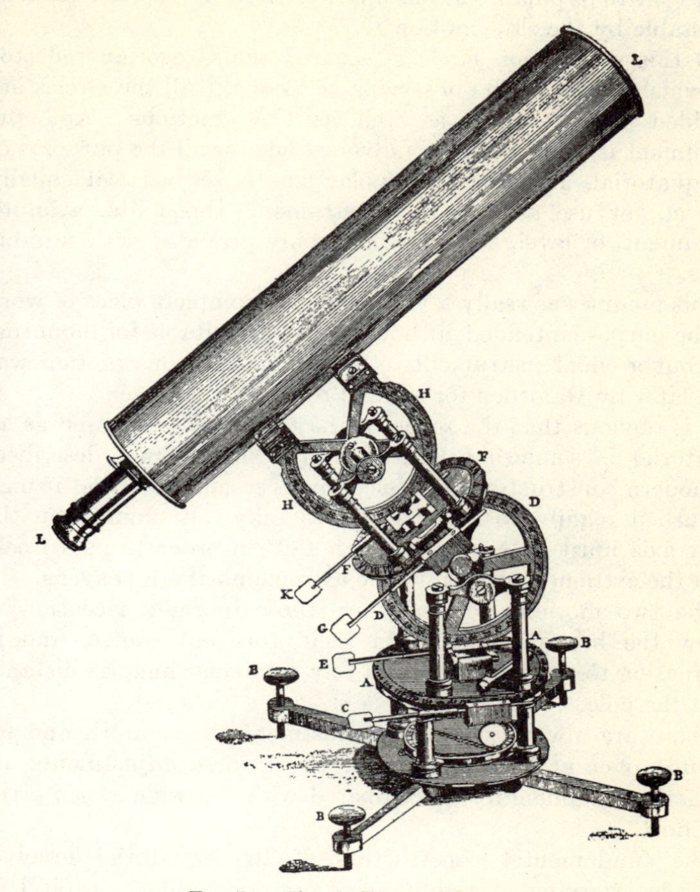

FIG. 74.—Short's Equatorial Mount.

In the instrument as shown there is first an azimuth circle *A A* supported on a base *B B B B* having levelling screws in the feet. Immediately under the azimuth circle is mounted a compass needle for approximate orientation, and the circle is adjustable by a tangent screw *C*.

Carried by the azimuth circle on a bearing supported by four pillars is a latitude circle *D D* for the adjustment of the polar axis, with a slow motion screw *E*. The latitude circle carries a right ascension circle *F F*, with a slow motion *G*, and this in turn carries on four pillars the declination circle *H H*, and its axis adjustable by the slow motion *K*.

To this declination circle is secured the Gregorian reflector *L L* which serves as the observing telescope. All the circles are provided with verniers as well as slow motions. And the instrument is, so to speak, a universal one for all the purposes of an equatorial, and when the polar axis is set vertical equally adapted for use as a transit instrument, theodolite, azimuth instrument, or level, since the circles are provided with suitable levels.

This mount was really a very neat and complete piece of work for the purpose intended, although scarcely suitable for mounting any but a small instrument. A very similar construction was used later by Ramsden for a small refractor.

It is obvious that the reach of the telescope when used as an equatorial is somewhat limited in the mount just described. In modern constructions the telescope is so mounted that it may be turned readily to any part of the sky, although often the polar axis must be swung through 180° in order to pass freely from the extreme southern to the extreme northern heavens.

The two motions necessary are those in right ascension to follow the heavenly bodies in their apparent course, and in declination to reach an object at any particular angular distance from the pole.

There are always provided adjustments in azimuth and for latitude over at least a small arc, but these adjustments are altogether rudimentary as compared with the wide sweep given by Short.

The fundamental construction of the equatorial involves two axes working at right angles positioned like a capital T.

The upright of the T is the polar axis, fitted to a sleeve and bearing the cross of the T, which is hollow and provides the bearing for the declination axis, which again carries at right angles to itself the tube of the telescope.

When the sleeve which carries the upright of the T points to the pole the telescope tube can evidently be swung to cover an object at any altitude, and can then be turned on its polar axis so as to follow that object in its apparent diurnal motion. The front fork of a bicycle set at the proper angle with a cross axis replacing the handle bars has more than once done good service

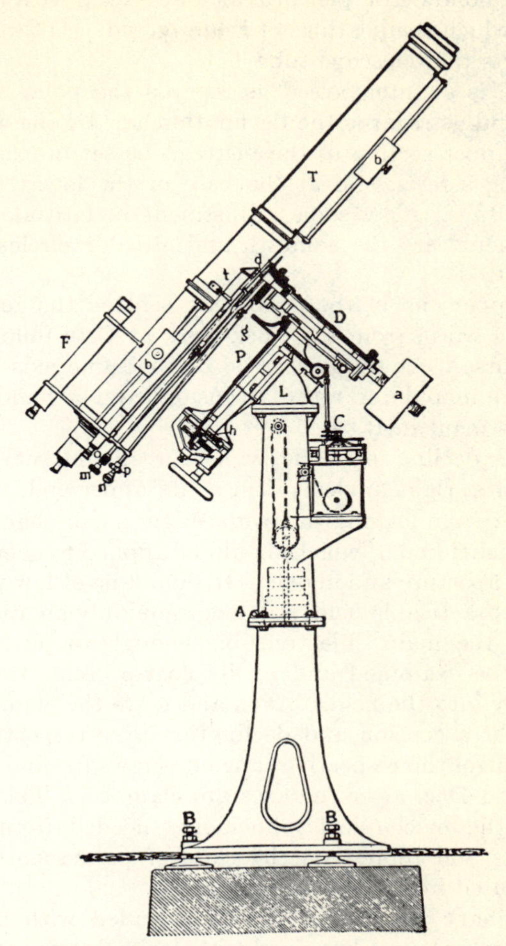

Fig. 75.—Section of Modern Equatorial.

in an emergency. Figure 75 shows in section a modern equatorial mount for a medium sized telescope.

The mounting shown in Fig. 75, by Zeiss, is thoroughly typical of recent practice in instruments of moderate size. The general form of this equatorial comes straight down to us from Fraunhofer's mounting of the Dorpat instrument. It consists essentially of two axes crossed exactly at right angles.

P, the polar axis, is aligned exactly with the pole, and is supported on a hollow iron pier provided at its top with the latitude block L to which the bearings of P are bolted. D the declination axis supports the telescope tube T.

The tube is counter-poised as regards the polar axis by the weight a, and as regards the declination axis by the weights b b. At A, the upper section of the pier can be set in exact azimuth by adjusting screws, and at the base of the lower section the screws at B. B. allow some adjustment in latitude. To such mere rudiments are the azimuth and altitude circles of Short's mount reduced.

At the upper end of the polar axis is fitted the gear wheel g, driven by a worm from the clock-work at C to follow the stars in their course. At the lower end of the same axis is the hour circle h, graduated for right ascension, and a hand wheel for quick adjustment in R. A.

At d is the declination circle, which is read, and set, by the telescope t with a right angled prism at its upper end, which saves the observer from leaving the eyepiece for small changes.

F is the usual finder, which should be applied to every telescope of 3 inches aperture and above. It should be of low power, with the largest practicable field, and has commonly an aperture $1/4$ or $1/5$ that of the main objective, big enough to pick up readily objects to be examined and by its coarse cross wires to bring them neatly into the field. At m and n are the clamping screws for the right ascension and declination axes respectively, while o and p control the respective tangent screws for fine adjustment in R. A. and Dec. after the axes are clamped. This mount has really all the mechanical refinements needed in much larger instruments and represents the class of permanently mounted telescopes used in a fixed observatory.

The ordinary small telescope is provided with a mount of the same general type but much simpler and, since it is not in a fixed observatory, has more liberal adjustments in azimuth and altitude to provide for changes of location. Figure 76 shows in some detail the admirable portable equatorial mounting used by the Clarks for instruments up to about 5 or 6 inches aperture.

Five inches is practically the dividing line between portable and fixed telescopes. In fact a 5-inch telescope of standard construction with equatorial mounting is actually too heavy for practical portability on a tripod stand. The Clarks have turned out really portable instruments of this aperture, of relatively short focus and with aluminum tube fitted to the mounting standard for a 4-inch telescope, but the ordinary 5-inch equipment of the usual focal length deserves a permanent placement.

In this mount the short tapered polar axis P is socketed between

FIG. 76.—Clark Adjustable Equatorial Mount.

the cheeks A, and tightened in any required position by the hand screws B. B. The stout declination axis D bears the telescope and the counterweight C. Setting circles in R. A. and Dec., p and d respectively, are carried on the two axes, and each axis has a worm wheel and tangent screw operated by a universal joint to give the necessary slow motion.

The worm wheels carry their respective axes through friction bearings and the counter poising is so exact that the instrument can be quickly swung to any part of the sky and the slow motion picked up on the instant. The wide sweep of the polar axis allows immediate conversion into an alt azimuth for terrestrial use, or adjustment for any latitude. A graduated latitude arc is customarily engraved on one of the check pieces to facilitate this adjustment.

Ordinarily portable equatorials on tripod mounts are not provided with circles, and have only a single slow motion, that in R. A. A declination circle, however, facilitates setting up the instrument accurately and is convenient for locating an object to be swept for in R. A. which must often be done if one has not sidereal time at hand. In Fig. 76 a thumb screw underneath the tripod head unclamps the mount so that it may be at once adjusted in azimuth without shifting the tripod.

As a rule American stands for fixed equatorials have the clock drive enclosed in the hollow pillar which carries the equatorial head as shown in the reflector of Fig. 35, and in the Clark mount for refractors of medium size shown in Fig. 77. Here a neat quadrangular pillar carries an equatorial mounting in principle very much like Fig. 76, but big enough to carry telescopes of 8 to 10 inches aperture. It has universal adjustment in latitude, so that it can be used in either hemisphere, the clock and its driving weight are enclosed in the pillar and slow motions are provided for finding in R. A. and Dec. The adjustment in azimuth is made by moving the pillar on its base-plate, which is bolted to the pier. The convenient connections for accurate following and the powerful clock make the mount especially good for photographic telescopes of moderate size and the whole equipment is most convenient and workmanlike. It is worth noting that the circles are provided with graduations that are plain rather than minute, in accordance with modern practice. In these days of celestial photography equatorials are seldom used for determining positions except with the micrometer, and graduated circles therefore, primarily used merely for finding, should be, above all things, easy to read.

All portable mounts are merely simplifications of the observatory type of Fig. 75, which, with the addition of various labor saving devices is applied to nearly all large refractors and to many reflectors as well.

There is a modified equatorial mount sometimes known as the "English" equatorial in which the polar axis is long and supported on two piers differing enough in height to give the proper latitude angle, the declination axis being about midway of the polar axis. A bit of the sky is cut off by the taller pier, and the type is not especially advantageous unless in supporting a very heavy instrument, too heavy to be readily overhung in the usual way.

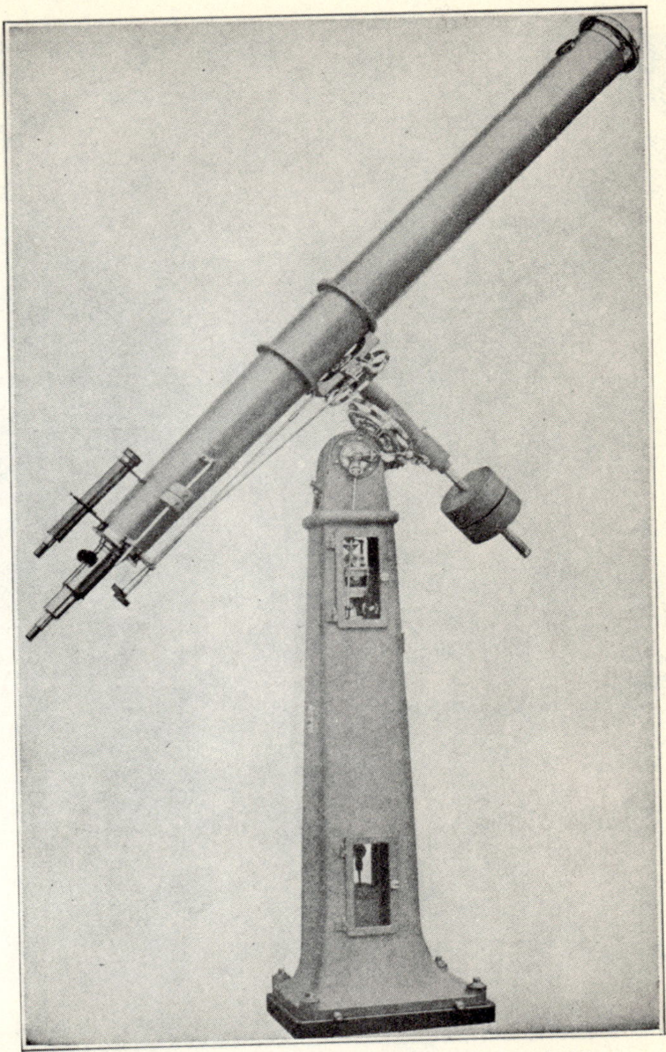

FIG. 77.—Universal Observatory Mount (Clark 9-inch).

In such case some form of the "English" mounting is very important to securing freedom from flexure and thereby the perfection of driving in R. A. so important to photographic work. The 72-inch Dominion Observatory reflector and the 100-inch Hooker telescope at Mt. Wilson are thus mounted, the former

FIG. 78.—English Equatorial Mount (Hooker 100-inch Telescope).

on a counterpoised declination axis crosswise the polar axis, the original "English" type; the latter on trunnions within a long closed fork which carries the polar bearings at its ends.

Figure 78 shows the latter instrument, of 100 inches clear aperture and of 42 feet principal focal length, increased to 135 feet when used as a Cassegrainian. It is the immense stability of this mount that has enabled it to carry the long cross girder bearing the interferometer recently used in measuring the diameters of

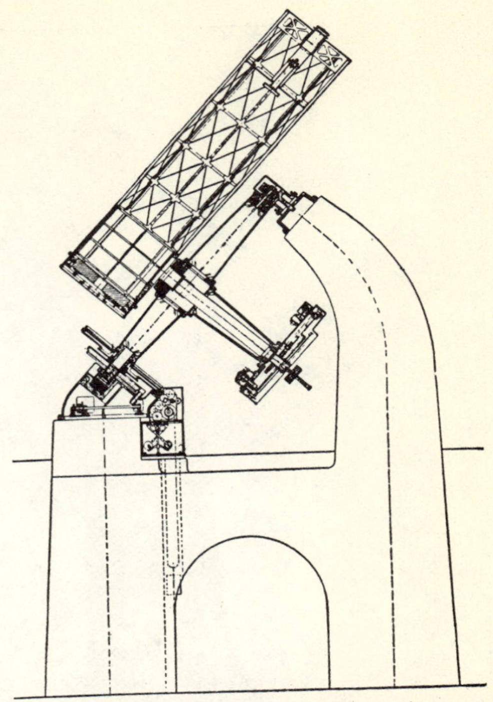

Fig. 79.—English Equatorial Mount (72-inch Dominion Observatory Telescope).

the stars. Note the mercury-flotation drum at each end of the polar axis. The mirrors were figured by the skillful hands of Mr. Ritchey.

Figure 79 gives in outline the proportions and mounting of the beautiful instrument in service at the Dominion Observatory, near Victoria, B. C. The mirror and its auxiliaries were figured by Brashear and the very elegant mounting was by Warner and Swasey. The main mirror is of 30 feet principal focus. The 20-inch hyperboloidal mirror extends the focus as a Cassagrainian to 108 feet. The mechanical stability of these English mounts for very large instruments has been amply demonstrated by this, as by the Hooker 100-inch reflector. They suffer less from flexure than the Fraunhofer mount where great weights are to be carried, although the latter is more convenient and generally useful for instruments of moderate size. It is hard to say too much of the mechanical skill that has made these two colossal

FIG. 80.—Astrographic Mount with Bent Pier.

telescopes so completely successful as instruments of research.

The inconvenience of having to swing the telescope tube to clear the pier at certain points in the R. A. following is often a serious nuisance in photographic work requiring long exposures, and may waste valuable time in visual work. Several recent forms of equatorial mount have therefore been devised to allow the telescope complete freedom of revolution in R. A., swinging clear of everything.

One such form is shown in Fig. 80 which is a standard astrographic mount for a Brashear doublet and guiding telescope. The pier is strongly overhung in the direction of the polar axis far enough to allow the instrument to follow through for any

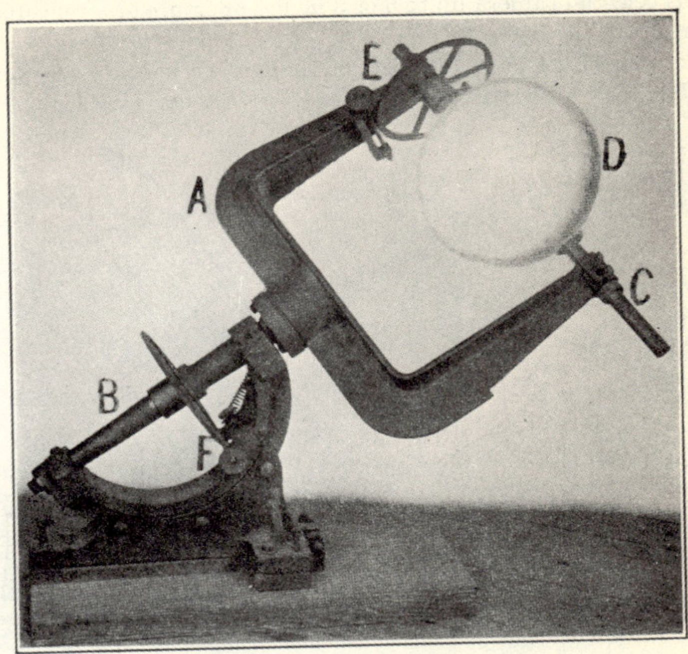

FIG. 81.—Open Fork Mounting.

required period, even to resuming operations on another night without a shift of working position.

Another form, even simpler and found to be extremely satisfactory even for rather large instruments, is the open polar fork mount. Here the polar axis of an ordinary form is continued by a wide and stiff casting in the form of a fork within which the tube is carried on substantial trunnions, giving it complete freedom of motion.

The open fork mount in its simplest form, carrying a heliostat mirror, is shown in Fig. 81. Here *A* is the fork, *B* the polar axis, carried on an adjustable sector for variation in latitude, *C* the declination axis carrying the mirror *D* in its cell, *E* the slow motion in declination, and F that in R. A. Both axes can be unclamped for quick motion and the R. A. axis can readily be driven by clock or electric motor.

The resemblance to the fully developed English equatorial mount of Fig. 78 is obvious, but the present arrangement gives entirely free swing to a short instrument, is conveniently adjustable, and altogether workmanlike. It can easily carry a short focus celestial camera up to 6 or 8 inches aperture or a reflector of 4 or 5 feet focal length.

In Fig. 173, Chap. X a pair of these same mounts are shown at Harvard Observatory. The nearer one, carrying a celestial

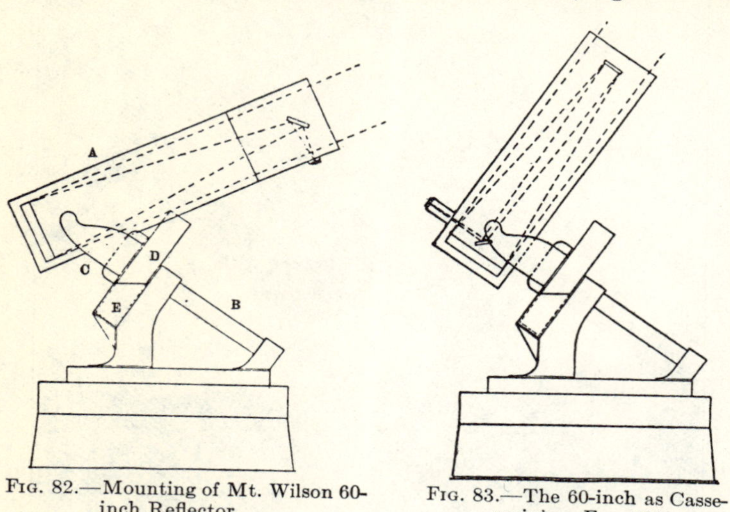

FIG. 82.—Mounting of Mt. Wilson 60-inch Reflector.

FIG. 83.—The 60-inch as Cassegrainian, F = 100′.

camera, is exposed to view. It is provided with a slow motion and clamp in declination, and with an electric drive in R. A., quickly unclamped for swinging the camera. It works very smoothly, its weight is taken by a very simple self adjusting thrust bearing at the lower end of the polar axis, and altogether it is about the simplest and cheapest equatorial mount of first class quality that can be devised for carrying instruments of moderate length.

Several others are in use at the Harvard Observatory and very similar ones of a larger growth carry the 24-inch Newtonian reflector there used for stellar photography and the 16-inch Metcalf photographic doublet.

In fact the open fork mount, which was developed by the late

Dr. Common, is very well suited to the mounting of big reflectors. It was first adapted by him to his 3 ft.-reflector and later used for his two 5 ft.-mirrors, and more recently for the 5 ft.-instrument at Mt. Wilson, and a good many others of recent make. Dr. Common in order to secure the easiest possible motion in R. A. devised the plan of floating most of the weight assumed by the polar axis in mercury.

Figure 82 is, diagrammatically, this fork mount as worked out by Ritchey for the 60-inch Mt. Wilson reflector. Here A is the lattice tube, B the polar axis, C the fork and D the hollow steel

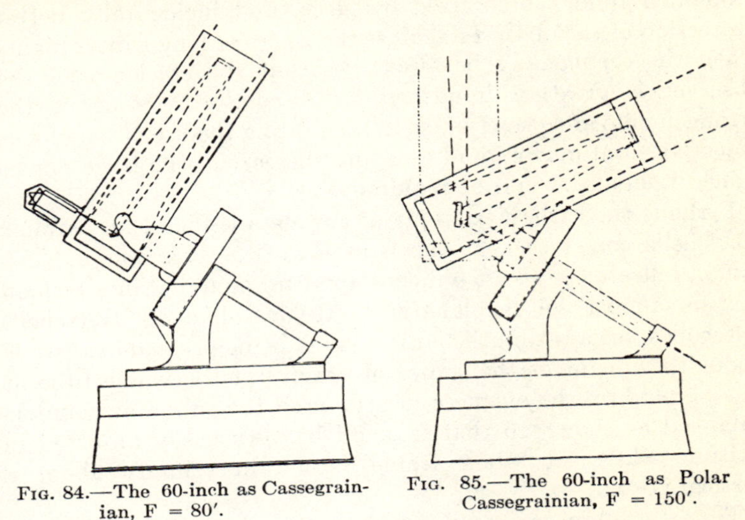

Fig. 84.—The 60-inch as Cassegrainian, F = 80′.

Fig. 85.—The 60-inch as Polar Cassegrainian, F = 150′.

drum which floats the axis in the mercury trough E. The great mirror is here shown worked as a simple Newtonian of 25 ft. focal length. As a matter of fact it is used much of the time as a Cassegranian.

To this end the upper section of tube carrying the oblique mirror is removed and a shorter tube carrying any one of three hyperboloidal mirrors is put in its place. Fig. 83 is the normal arrangement for visual or photographic work on the long focus, 100 ft. The dotted lines show the path of the rays and it will be noticed that the great mirror is not perforated as in the usual Cassegrainian construction, but that the rays are brought out by a diagonal flat.

Figure 84 is a similar arrangement used for stellar spectroscopy with a small flat and an equivalent focus of 80 ft. In Fig. 85 a radically different scheme is carried out. The hyperboloidal mirror now used gives an equivalent focus of 150 ft., and the auxiliary flat is arranged to turn on an axis parallel to the declination axis so as to send the reflected beam down the hollow polar axis into a spectrograph vault below the southern end of the axis. Obviously one cannot work near the pole with this arrangement but only through some 75° as indicated by the dotted lines. · The fork mount is not at all universal for reflectors, as has already been seen, and Cassegrainiaus of moderate size are very commonly mounted exactly like refractors.

We now come to a group of mounts which have in common the fundamental idea of a fixed eyepiece, and incidentally better protection of the observer against the rigors of long winter nights when the seeing may be at its best but the efficiency of the observer is greatly diminished by discomfort. Some of the arrangements are also of value in facilitating the use of long focus objectives and mirrors and escaping the cost of the large domes which otherwise would be required.

Perhaps the earliest example of the class is found in Caroline Herschel's comet seeker, shown in Fig. 86. This was a Newtonian reflector of about 6 inches aperture mounted in a fashion that is almost self explanatory. It was, like all Herschel's telescopes, an altazimuth but instead of being pivoted in altitude about the mirror or the center of gravity of the whole tube, it was pivoted on the eyepiece location and the tube was counterbalanced as shown so that it could be very easily adjusted in altitude while the whole frame turned in azimuth about a vertical post.

Thus the observer could stand or sit at ease sweeping in a vertical circle, and merely had to move around the post as the azimuth was changed. The arrangement is not without advantages, and was many years later adopted with modifications of detail by Dr. J. W. Draper for the famous instrument with which he advanced so notably the art of celestial photography.

The same fundamental idea of freeing the observer from continual climbing about to reach the eyepiece has been carried out in various equatorially mounted comet seekers. A very good example of the type is a big comet seeker by Zeiss, shown in Fig. 87. The fundamental principle is that the ocular is at the intersection of the polar and declination axis, the telescope tube being overhung well beyond the north end of the former and counterbalanced on the latter. The observer can therefore sit in his swivel chair and without stirring from it sweep rapidly over a very wide expanse of sky.

This particular instrument is probably the largest of regular comet seekers, 8 inches in clear aperture and 52½ inches focal length with a triple objective to ensure the necessary corrections in using so great a relative aperture. In this figure 1 is the base with corrections in altitude and azimuth, 2 the counterpoise of the whole telescope on its base, 3 the polar axis and R. A. circle, 4 the overhung declination axis and its circle, 5 the counterpoise in declination, 6 the polar counterpoise, and 7 the main telescope tube. The handwheel shown merely operates the gear for revolving the dome without leaving the observing chair.

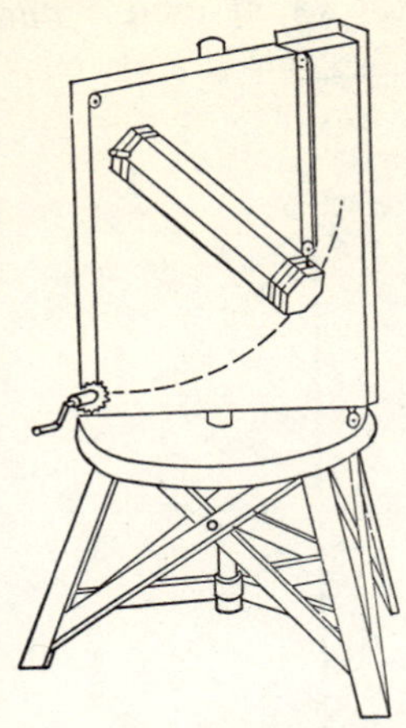

Fig. 86.—Caroline Herschel's Comet Seeker.

The next step beyond the eyepiece fixed in general position is to locate it so that the observer can be thoroughly protected without including the optical parts of the telescope in such wise as to injure their performance.

One cannot observe successfully through an open window on account of the air currents due to temperature differences, and in an observatory dome, unheated as it is, must wait after the shutter is opened until the temperature is fairly steadied.

Except for these comet seekers practically all of the class make use of one or two auxiliary reflections to bring the image into the required direction, and in general the field of possible view is somewhat curtailed by the mounting. This is less of a disadvantage than it would appear at first thought, for, to begin with, observations within 20° of the horizon or thereabouts are generally unsatisfactory, and the advantages of a stable and convenient long focus instrument are so notable as for many purposes quite to outweigh some loss of sky-space.

The simplest of the fixed eyepiece group is the polar telescope of which the rudiments are well shown in Fig. 88, a mount described by Sir Howard Grubb in 1880, and an example of which was installed a little later in the Crawford Observatory in Cork. Here the polar axis A is the main tube of the telescope, and in front of the objective B, is held in a fork the declination cradle and mirror C, by which any object within a wide sweep of declination can be brought into the field and held there by hand or clockwork through rotating the polar tube.

Looked at from another slant it is a polar heliostat, of which the telescope forms the driving axis in R. A. The whole mount was

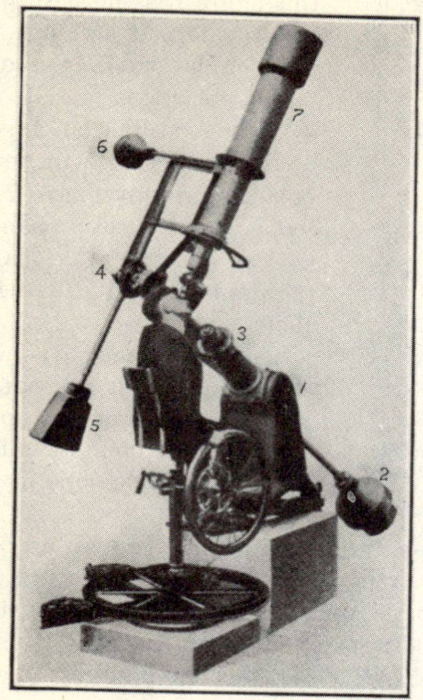

FIG. 87.—Mounting of Large Comet Seeker.

a substantial casting on wheels which ran on a pair of rails. For use the instrument was rolled to a specially arranged window and through it until over its regular bearings on a pier just outside.

A few turns of the wheel D lowered it upon these, and the back of the frame then closed the opening in the wall leaving the instrument in the open, and the eye end inside the room. The example first built was of only 4 inches aperture but proved its case admirably as a most useful and convenient instrument.

This mount with various others of the fixed eyepiece class may be regarded as derived from the horizontal photoheliographs used at the 1874 transit of Venus and subsequently at many total solar eclipses. Given an equatorially mounted heliostat like Fig. 81 and it is evident that the beam from it may be turned into a horizontal telescope placed in the meridian, (or for that matter in any convenient direction) and held there by rotation of

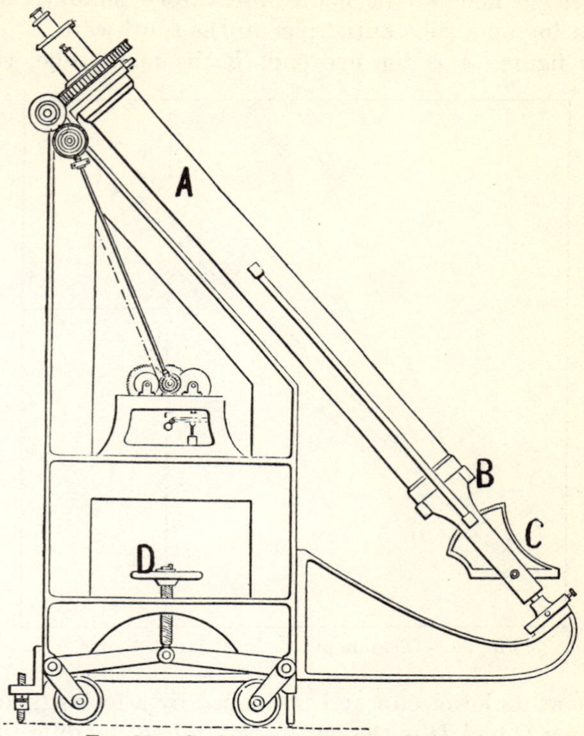

Fig. 88.—Grubb's Original Polar Telescope.

the mirror in R. A., but also in declination, save in the case where the beam travels along the extension of the polar axis.

For the brief exposure periods originally needed and the slow variation of the sun in declination this heliostatic telescope was easily kept in adjustment. The original instruments were of 5 inches aperture and 40 ft. focal length, and the 7-inch heliostat mirror was provided with ordinary equatorial clockwork. Set up with the telescope pointing along the polar axis no continuous variation in declination is needed and the clock drive holds the field steadily, as in any other equatorial.

Figure 89 shows diagrammatically the 12-inch polar telescope used for more than twenty years past at the Harvard Observatory. The mount was designed by Mr. W. P. Gerrish of the Harvard staff and contains many ingenious features. Unlike Fig. 88 this is a fixed mount, with the eye-end comfortably housed in a room on the second floor of the main observatory building, and the lower bearing on a substantial pier to the southward.

In the figure, *A* is the eye end, *B* the main tube with the

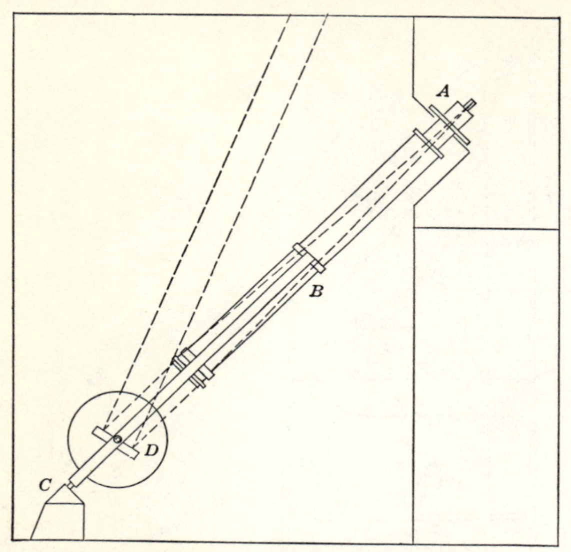

FIG. 89.—Diagram of Gerrish Polar Telescope.

objective at its lower end and prolonged by a fork supported by the bearing *C* and *D* is the declination mirror sending the beam upward. The whole is rotated in R. A. by an electric clock drive, and all the necessary adjustments are made from the eye end.

A view of the exterior is shown in Fig. 90, with the mirror and objective uncovered. The rocking arm at the objective end, operated by a small winch beside the ocular, swings clear both mirror and objective caps in a few seconds, and the telescope is then ready for use. Its focal length is 16 ft. 10 inches and it gives a sweep in declination of approximately 80°. It gives excellent definition and has proved a most useful instrument.

A second polar telescope was set up at the Harvard Observatory station in Mandeville, Jamaica, in the autumn of 1900.

This was intended primarily for lunar photography and was provided with a 12-inch objective of 135 ft. 4 inches focal length and an 18-inch heliostat with electric clock drive.

Inasmuch as all instruments of this class necessarily rotate

Fig. 90.—Gerrish Polar Telescope, Harvard Observatory.

the image as the mirror turns, the tail-piece of this telescope is also mounted for rotation by a similar drive so that the image is stationary on the plate both in position and orientation. As Mandeville is in N. lat. 18° 01′ the telescope is conveniently near the horizontal. The observatory of Yale University has a large instrument of this class, of 50 feet focal length, with a 15-inch photographic objective and a 10-inch visual guiding objective working together from the same heliostat.

Despite its simplicity and convenience the polar telescope has an obvious defect in its very modest sweep in declination, only to be increased by the use of an exceptionally large mirror. It is not therefore remarkable that the first serious attempt at a fixed eyepiece instrument for general use turned to a different construction even at the cost of an additional reflection.

This was the *equatorial coudé* devised by M. Loewy of the Paris Observatory in 1882. (Fig. 91.) In the diagram A is the main tube which forms the polar axis, and B the eye end under shelter, with all accessories at the observers hand. But the tube is broken by the box casing C containing a mirror rigidly supported at 45° to the axis of the main tube and of the side tube D, which is counterbalanced and is in effect a hollow declination axis carrying the objective E at its outer end.

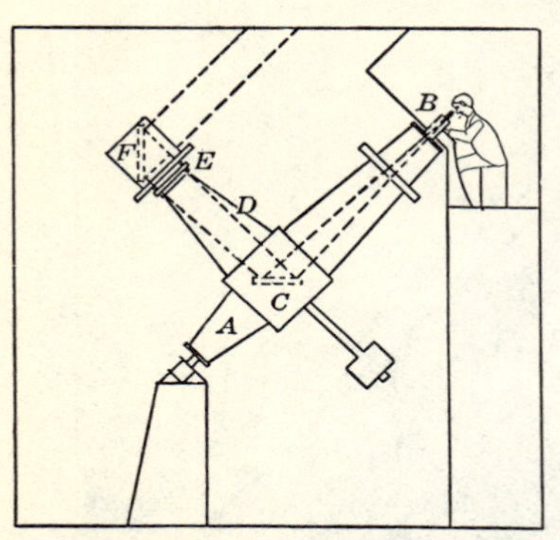

FIG. 91.—Diagram of Equatorial Coudé.

In lieu of the telescope tube usually carried on this declination axis we have the 45° mirror, F, turning in a sleeve concentric with the objective, which, having a lateral aperture, virtually gives the objectives a full sweep in declination, save as the upper pier cuts it off. The whole instrument is clock driven in R. A., and has the usual circles and slow motions all handily manipulated from the eye end.

The *equatorial coudé* is undeniably complicated and costly, but as constructed by Henry Frères it actually performs admirably even under severe tests, and has been several times duplicated in French observatories. The first *coudé* erected was of 10½ inches aperture and was soon followed by one of 23.6 inches aperture and 59 ft. focus, which is the largest yet built.

Still another mounting suggestive of both the polar telescope and the *coudé* is due to Sir Howard Grubb, Fig. 92. Here as in the *coudé* the upper part of the polar axis, *A*, is the telescope tube which leads into a solid casing *B*, about which a substantial fork, *C*, is pivoted. This fork is the extension of the side tube *D*, which carries the objective, and thus has free swing in declination through an angle limited by the roof of the observing room above, and the proximity of the horizon below.

Its useful swing, as in the polar telescope, is limited by the dimensions of the mirror *E*, which receives the cone of rays from the objective and turn it up the polar tube to the eyepiece. This mirror is geared to turn at half the rate of the tube *D* so that the angle *D E A* is continually bisected.

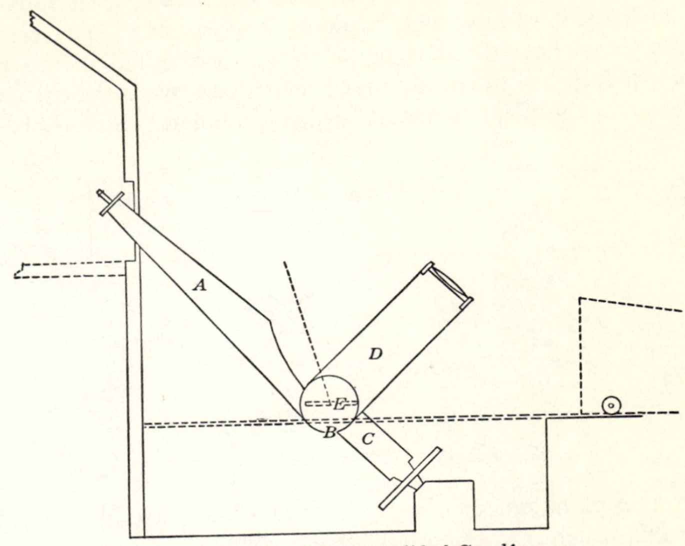

Fig. 92.—Grubb Modified Coudé.

In point of fact the sole gain in this construction is the reduction in the size of mirror required, by reason of the diminished size of the cone of rays when it reaches the mirror. The plan has been very successfully worked out in the fine astrographic telescope of the Cambridge Observatory of 12½ inches aperture and 19.3 ft. focal length.

As in the other instruments of this general class the adjustments are all conveniently made from the eye end. The Cambridge instrument has a triple photo-visual objective of the form designed by Mr. H. D. Taylor and the side tube, when not in use, is turned down to the horizontal and covered in by a low wheeled housing carried on a track. The sky space covered is from 15° above the pole to near the horizontal.

It is obvious that various polar and *coudé* forms of reflector are quite practicable and indeed one such arrangement is shown in connection with the 60-inch Mt. Wilson reflector, but we are here concerned only with the chief types of mounting which have actually proved their usefulness. None of the arrangements which require the use of additional large reflecting surfaces are exempt from danger of impaired definition. Only superlatively fine workmanship and skill in mounting can save them from distortion and astigmatism due to flexure and warping of the mirrors, and such troubles have not infrequently been encountered.

To a somewhat variant type belong several valuable constructions which utilize in the auxiliary reflecting system the coelostat rather than the polar heliostat or its equivalent. The coelostat

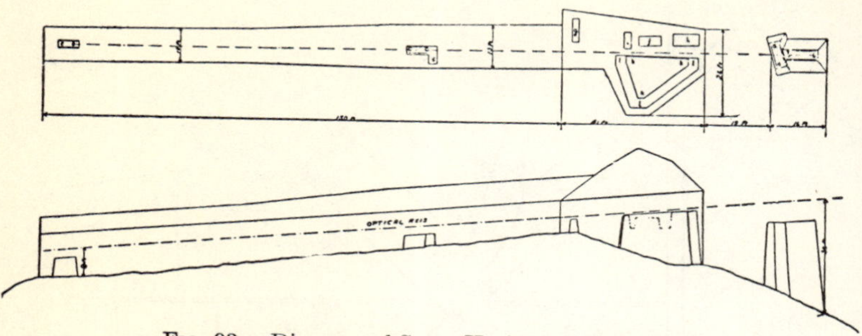

Fig. 93.—Diagram of Snow Horizontal Telescope.

is simply a plane mirror mounted with its plane fixed in that of a polar axis which rotates once in 48 hours, i.e., at half the apparent rate of the stars.

A telescope pointed at such a mirror will hold the stars motionless in its field as if the firmament were halted á la Joshua. But if a change of view is wanted the telescope must be shifted in altitude or azimuth or both. This is altogether inconvenient, so that as a matter of practice a second plane mirror is used to turn the steady beam from the coelostat into any desired direction.

By thus shifting the mirror instead of the telescope, the latter can be permanently fixed in the most convenient location, at the cost of some added expense and loss of light. Further, the image does not rotate as in case of the polar heliostat, which is often an advantage.

An admirable type of the fixed telescope thus constituted is the Snow telescope at Mt. Wilson (Cont. from the Solar Obs.

No. 2, Hale). Fig. 93 from this paper shows the equipment in plan and elevation. The topography of the mountain top made it desirable to lay out the axis of the building 15° E. of N. and sloping downward 5° toward the N.

At the right hand end of the figure is shown the coelostat pier, 29 ft. high at its S end. This pier carries the cœlostat mirror proper, 30 inches in diameter, on rails *a a* accurately E. and W. to allow for sliding the instrument so that its field may clear the secondary mirror of 24 inches diameter which is on an altazimuth fork mounting and also slides on rails *b b*.

The telescope here is a pair of parabolic mirrors each of 24 inches aperture and of 60 ft. and 145 ft. focus respectively. The beam from the secondary cœlostat mirror passes first through the spectrographic laboratory shown to the left of the main pier, and in through a long and narrow shelter house to one of these mirrors; the one of longest focus on longitudinal focussing rails *e e*, the other on similar rails *c c*, with provision for sliding sidewise at *d* to clear the way for the longer beam.

The ocular end of this remarkable telescope is the spectrographic laboratory where the beam can be turned into the permanently mounted instruments, for the details of which the original paper should be consulted. The purpose of this brief description is merely to show the beautiful facility with which a cœlostatic telescope may be adapted to astrophysical work. Obviously an objective could be put in the cœlostat beam for any purpose for which it might be desirable.

Such in fact is the arrangement of the tower telescopes at the Mt. Wilson Observatory. In these instruments we have the ordinary cœlostat arrangement turned on end for the sake of getting the chief optical parts well above the ground where, removed from the heated surface, the definition is generally improved. To be sure the focus is at or near the ground level, but the upward air currents cause much less disturbance than the crosswise ones in the Snow telescope.

The head of the first tower telescope is shown in Fig. 94.* A is the coelostat mirror proper 17 inches in diameter and 12 inches thick, B the secondary mirror 12¾ inches in the shorter axis of the ellipse, 22¼ inches in the longer, and also 12 inches

* Contributions from the Solar Obs. No. 23, Hale, which should be seen for details.

thick. C is the 12-inch objective of 60 ft. focus, and D the focussing gear worked by a steel ribbon from below.

This instrument being for solar research the mirrors are

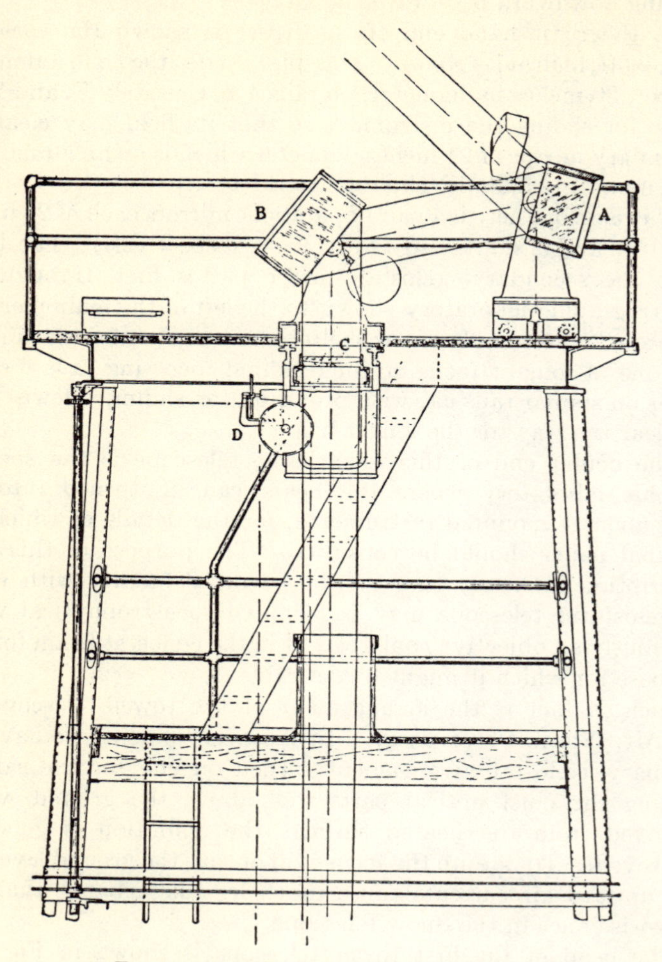

FIG. 94.—Head of 60-foot Tower Telescope.

arranged for convenient working with the sun fairly low on either horizon where the definition is at its best, and can be shifted accordingly, to the same end as in the Snow telescope. There is also provision for shifting the objective laterally at a uniform rate from below, to provide for the use of the apparatus as spectro-heliograph.

The tower is of the windmill type and proved to be fairly steady in spite of its height, high winds being rare on Mt. Wilson. The great thickness of the mirrors in the effort to escape distortion deserves notice. They actually proved to be too

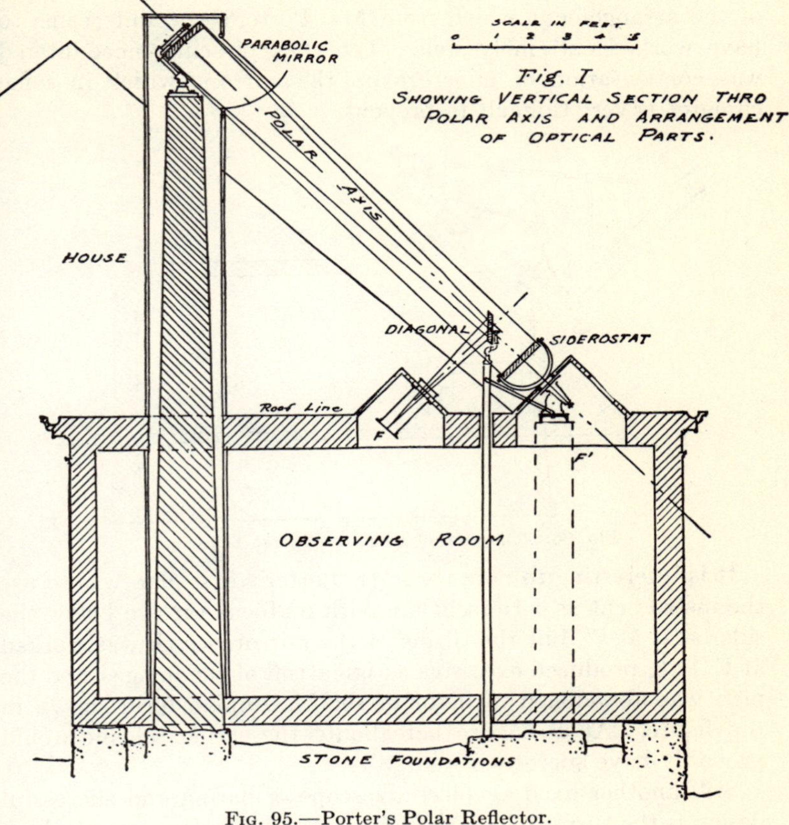

Fig. 95.—Porter's Polar Reflector.

thick to give thermal conductivity sufficient to prevent distortion.

In the later 150-foot tower telescope the mirrors are relatively less thick, and a very interesting modification has been introduced in the tower, in that it consists of a lattice member for member within another exterior lattice, so that the open structure is retained, while each member that supports the optical parts is shielded from the wind and sudden temperature change by its corresponding outer sheath.

Still another form of mounting to give the observer access to a fixed eyepiece under shelter is found in the ingenious polar reflector by Mr. Russell W. Porter of which an example with main mirror of 16 inches diameter and 15 ft. 6 inches focal length was erected by him a few years ago. Fig. 95 is entirely descriptive of the arrangement which from Mr. Porter's account seems to have worked extremely well. The chief difficulty encountered was condensation of moisture on the mirrors, which in some climates is very difficult to prevent.

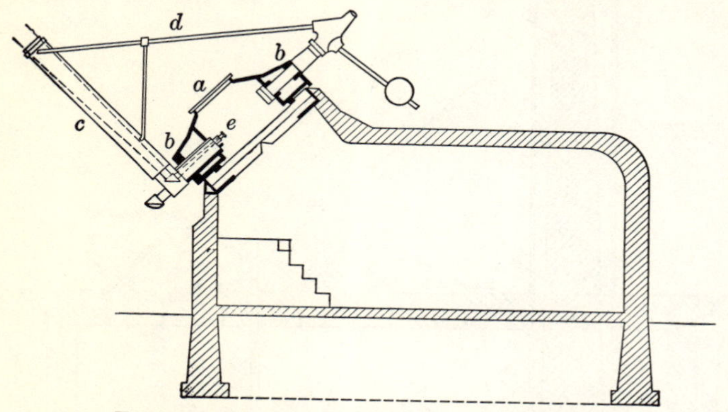

Fig. 96.—Diagram of Hartness Turret Telescope.

It is interesting to note that Mr. Porter's first plan was to use the instrument as a Herschelian with its focus thrown below the siderostat at F', but the tilting of the mirror, which was worked at F/11.6, produced excessive astigmatism of the images, and the plan was abandoned in favor of the Newtonian form shown in the figure. At F/25 or thereabouts the earlier scheme would probably have succeeded well.

Still another fixed eyepiece telescope of daring and successful design is the turret telescope of the Hon. J. E. Hartness of which the inventor erected a fine example of 10 inches aperture at Springfield, Vermont. The telescope is in this case a refractor, and the feature of the mount is that the polar axis is expanded into a turret within which the observer sits comfortably, looking into the ocular which lies in the divided declination axis and is supplied from a reflecting prism in the main beam from the objective.

Figure 96 shows a diagram of the mount and observatory. Here *a* is the polar turret, *bb* the bearings of the declination axis, *c* the main tube, *d* its support, and *e* the ocular end. Optically the telescope is merely an ordinary refractor used with a right angled prism a little larger and further up the tube than usual. The turret is entered in this instance from below, through a tunnel from the inventor's residence. The telescope as shown in Fig. 96 has a 10-inch Brashear objective of fine optical quality, and the light is turned into the ocular tube by a right angled prism only 2¾ inches in the face. This is an entirely practicable size for a reflecting prism and the light lost is not materially in excess of that lost in the ordinary "star diagonal" so necessary for the observation of stars near the zenith in an ordinary equatorial. The only obvious difficulty of the construction is the support of the very large polar axis. Being an accomplished mechanical engineer, Mr. Hartness worked out the details of this design very successfully although the moving parts weighed some 2 tons. The ocular is not absolutely fixed with reference to the observer but is always conveniently placed, and the performance of the instrument is reported as excellent in every respect, while the sheltering of the observer from the rigors of a Vermont winter is altogether admirable. Figure 97 shows the complete observatory as it stands. Obviously the higher the latitude the easier is this particular construction, which lends itself readily to large instruments and has the additional advantage of freeing the observer from the insect pests which are extremely troublesome in warm weather over a large part of the world.

This running account of mountings makes no claim at completeness. It merely shows the devices in common use and some which point the way to further progress. The main requirements in a mount are steadiness, and smoothness of motion. Even an altazimuth mount with its need of two motions, if smooth working and steady, is preferable to a shaky and jerky equatorial.

Remember that the Herschels did immortal work without equatorial mountings, and used high powers at that. A clock driven equatorial is a great convenience and practically indispensable for the photographic work that makes so large a part of modern astronomy, but for eye observations one gets on very fairly without the clock.

Circles are a necessity in all but the small telescopes used on portable tripods, otherwise much time will be wasted in finding.

FIG. 97.—Hartness Turret Observatory from the N. E.

In any event do not skimp on the finder, which should be of ample aperture and wide field, say ¼ the aperture of the main objective, and 3° to 5° in field. Superior definition is needless, light, and sky room enough to locate objects quickly being the fundamental requisites.

As a final word see that all the adjustments are within easy reach from the eyepiece, since an object once lost from a high power ocular often proves troublesome to locate again.

REFERENCES

Chambers' Astronomy, Vol. II.

F. L. O. Wadsworth: *Ap. J.*, **5**, 132. Ranyard's mounts for reflectors.

G. W. Ritchey: *Ap. J.*, **5**, 143. Supporting large specula.

G. E. Hale: Cont. Mt. W. Obs. No. 2. The "Snow" horizontal telescope.

G. E. Hale: Cont. Mt. W. Obs. No. 23. The 60 ft. tower telescope.

J. W. Draper: Smithsonian Contrib. **34.** Mounting of his large reflector.

G. W. Ritchey: Smithsonian Contrib. **35.** Mounting of the Mt. Wilson 60-inch reflector.

Sir H. Grubb: *Tr.* Roy. Dublin Soc. Ser. 2. **3.** Polar Telescopes.

Sir R. S. Ball: *M. N.* **59,** 152. Photographic polar telescope.

A. A. Common: Mem. R. A. S., **46,** 173. Mounting of his 3 ft. reflector.

R. W. Porter: *Pop. Ast.*, **24,** 308. Polar reflecting telescope.

James Hartness: *Trans. A. S. M. E.*, 1911, Turret Telescope.

Sir David Gill: Enc. Brit., 11th Ed. Telescope. Admirable summary of mounts.

CHAPTER VI

EYEPIECES

The eyepiece of a telescope is merely an instrument for magnifying the image produced by the objective or mirror. If one looks through a telescope without its eyepiece, drawing the eye back from the focus to its ordinary distance of distinct vision, the image is clearly seen as if suspended in air, or it can be received on a bit of ground glass.

It appears larger or smaller than the object seen by the naked eye, in proportion as the focal length of the objective is larger or smaller than the distance to which the eye has to drop back to see the image clearly.

This real image, the quality of which depends on the exactness of correction of the objective or mirror, is then to be magnified so much as may be desirable, by the eyepiece of the instrument. In broad terms, then, the eyepiece is a simple microscope applied to the image of an object instead of the object itself.

And looking at the matter in the simplest way the magnifying power of any simple lens depends on the focal length of that lens compared with the ordinary seeing distance of the eye. If this be taken at 10 inches as it often conventionally is, then a lens of 1 inch focus brings clear vision down to an inch from the object, increases the apparent angle covered by the object 10 times and hence gives a magnifying power of 10.

But if the objective has a focal length of 100 inches the image, as we have just seen, is already magnified 10 times as the naked eye sees it, hence with an objective of 100 inches focus and a 1-inch eyepiece the total magnification is 100 diameters. And this expresses the general law, for if we took the normal seeing distance of the naked eye at some other value than 10 inches, say 12½ inches then we should have to reckon the image as magnified by 8 times so far as the objective inches is concerned, but 12½ times due to the 1-inch eyepiece, and so forth. Thus the magnifying power of any eyepiece is F/f where F is the focal length of the objective or mirror and f that of the eyepiece. The focal distance of the eye quite drops out of the reckoning.

All these facts appear very quickly if one explores the image from an objective with a slip of ground glass and a pocket lens. An ordinary camera tells the same story. A distant object which covers 1° will cover on the ground glass 1° reckoned on a radius equal to the focal length of the lens. If this is equal to the ordinary distance of clear vision, an eye at the same distance will see the image (or the distant object) covering the same 1°.

The geometry of the situation is as follows: Let *o* Fig. 5, Chap. I, be the objective. This lens, as in an ordinary camera, forms an inverted image of an object A B at its focus, as at *a b*, and for any point, as *a*, of the image there is a corresponding point of the object lying on the straight line from A to that point through the center, *c*, of the objective.

A pair of rays 1, 2, diverging from the object point A pass through rim and center of *o* respectively and meet in *a*. After crossing at this point they fall on the eye lens *e*, and if *a* is nearly in the principal focus of *e*, the rays 1 and 2 will emerge substantially parallel so that the eye will unite them to form a clear image.

Now if F is the focal length of *o*, and f that of *e*, the object forming the image subtends at the center of the objective, o, an angle *A c B*, and for a distant object this will be sensibly the angle under which the eye sees the same object.

If L is the half linear dimension of the image, the eye sees half the object covering the angle whose tangent is L/F. Similarly half the image *ab* is seen through the eye lens *e* as covering a half angle whose tangent is L/f. Since the magnifying power of the combination, m, is directly as the ratio of increase in this tangent of the visual angle, which measures the image dimension

$$m = \frac{F}{f}, \text{ as before}$$

Further, as all the light which comes in parallel through the whole opening of the objective forms a single conical beam concentrating into a focus and then diverging to enter the eye lens, the diameter of the cone coming through the eye lens must bear the same relation to the diameter of *o*, that f does to F.

Any less diameter of *e* will cut off part of the emerging light; any more will show an emergent beam smaller than the eye lens, which is generally the case. Hence if we call p the diameter of the bright pencil of light which we see coming through the eye lens then for that particular eye lens,

$$m = \frac{o}{p}$$

That is, $f = \frac{pF}{o}$, which is quite the easiest way of measuring the focal length of an eyepiece.

Point the telescope toward the clear sky, focusing for a distant object so that the emergent pencil is sharply defined at the ocular, and then measure its diameter by the help of a fine scale and a pocket lens, taking care that scale and emergent pencil are simultaneously in sharp focus and show no parallax as the eye is shifted a bit. This bright circle of the emerging beam is actually the projection by the eye lens of the focal image of the objective aperture.

This method of measuring power is easy and rather accurate. But it leads to trouble if the measured diameter of the objective is in fact contracted by a stop anywhere along the path of the beam, as occasionally happens. Examine the telescope carefully with reference to this point before thus testing the power.*

The eye lens of Fig. 5 is a simple double convex one, such as was used by Christopher Scheiner and his contemporaries. With a first class objective or mirror the simple eye lens such as is shown in Fig. 98*a* is by no means to be despised even now. Sir William Herschel always preferred it for high powers, and speaks with evident contempt of observers who sacrificed its advantages to gain a bigger field of view. Let us try to fathom the reason for his vigorously expressed opinion, strongly backed up by experienced observers like the late T. W. Webb and Mr. W. F. Denning.

First of all a single lens saves about 10% of the light. Each surface of glass through which light passes transmits 95 to 96 % of that light, so that a single lens transmits approximately 90%, two lenses 81% and so on. This loss may be enough to determine the visibility of an object. Sir Wm. Herschel found that faint objects invisible with the ordinary two-lens eyepiece came to view with the single lens.

* A more precise method, depending on an actual measurement of the angle subtended by the diameter of the eyepiece diaphragm as seen through the eye end of the ocular and its comparison with the same angular diameter reckoned from the objective, is given by Schaeberle. M. N. **43,** 297.

Probably the actual loss is less serious than its effect on seeing conditions. The loss is due substantially to reflection at the surfaces, and the light thus reflected is scattered close to, or into, the eye and produces stray light in the field which injures the contrast by which faint objects become visible.

In some eyepieces the form of the surfaces is such that reflected light is strongly concentrated where the eye sees it, forming a "ghost" quite bright enough greatly to interfere with the vision of delicate contrasts.

The single lens has a very small sharp field, hardly 10° in angular extent, the image falling off rapidly in quality as it departs from the axis. If plano-convex, as is the eye lens of

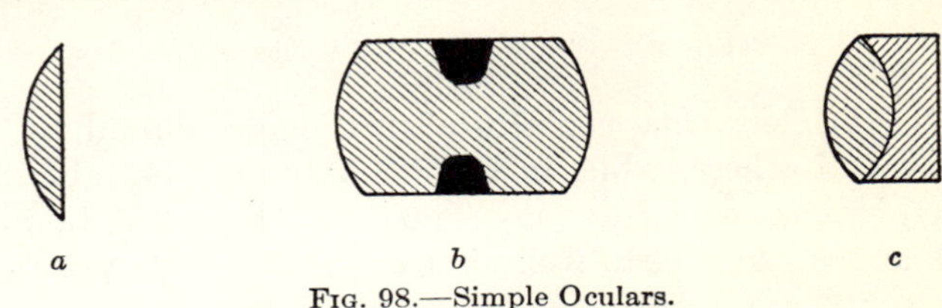

FIG. 98.—Simple Oculars.

common two-lens oculars, it works best with the curved side to the eye, i.e., reversed from its usual position, the spherical aberration being much less in this position.

Herschel's report of better definition with a single lens than with an ordinary two-lens ocular speaks ill for the quality of the latter then available. Of course the single lens gives some chromatic aberration, generally of small account with the narrow pencils of light used in high powers.

A somewhat better form of eye lens occasionally used is the so-called Coddington lens, really devised by Sir David Brewster. This, Fig. 98*b*, is derived from a glass sphere with a thick equatorial belt removed and a groove cut down centrally leaving a diameter of less than half the radius of the sphere. The focus is, for ordinary crown glass, $\frac{3}{2}$ the radius of the sphere, and the field is a little improved over the simple lens, but it falls off rather rapidly, with considerable color toward the edge.

The obvious step toward fuller correction of the aberrations while retaining the advantages of the simple lens is to make the ocular achromatic, like a minute objective, thus correcting at once the chromatic and spherical aberrations over a reasonably large field. As the components are cemented the loss of light at their common surface is negligible. Figure 98*c* shows such a lens. If correctly designed it gives an admirably sharp field of 15° to 20°, colorless and with very little distortion, and is well adapted for high powers.

Still better results in field and orthoscopy can be attained by going to a triple cemented lens, similar to the objective of

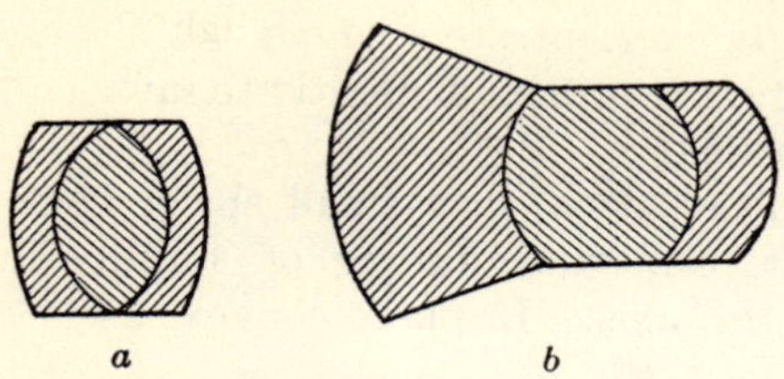

Fig. 99.—Triple Cemented Oculars.

Fig. 57. Triplets thus constituted are made abroad by Zeiss, Steinheil and others, while in this country an admirable triplet designed by Professor Hastings is made by Bausch & Lomb.

Such lenses give a beautifully flat and sharp field over an angle of 20° to 30°, quite colorless and orthoscopic. Fig. 99*a*, a form

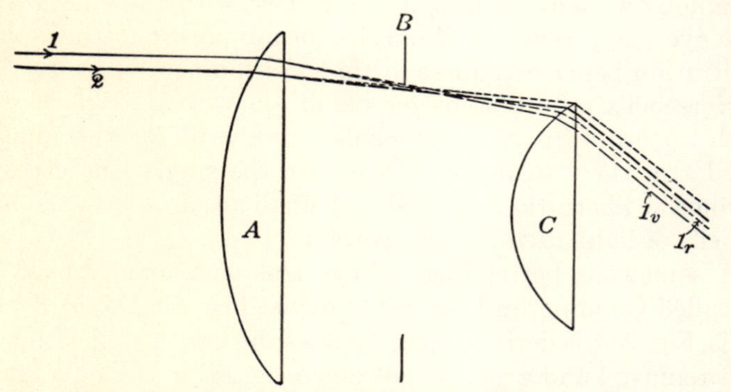

Fig. 100.—Path of Rays Through Huygenian Ocular.

used by Steinheil, is an excellent example of the construction and a most useful ocular. The late R. B. Tolles made such triplets, even down to ⅛ inch focus, which gave admirable results.

A highly specialized form of triplet is the so-called monocentric of Steinheil Fig. 99*b*. Its peculiarity is less in the fact that all the curves are struck from the same center than in the great thickness of the front flint and the crown, which, as in some photographic lenses, give added facilities for flattening the field and eliminating distortion.

The monocentric eyepiece has a high reputation for keen definition and is admirably achromatic and orthoscopic. The sharp field is about 32°, rather the largest given by any of the cemented combinations. All these optically single lenses are quite free of ghosts, reduce scattered light to a minimum, and leave little to be desired in precise definition. The weak point of the whole tribe is the small field, which, despite Herschel's opinion, is a real disadvantage for certain kinds of work and wastes the observer's time unless his facilities for close setting are more than usually good.

Hence the general use of oculars of the two-lens types, all of them giving relatively wide fields, some of them faultless also in definition and orthoscopy. The earliest form, Fig. 100, is the very useful and common one used by Huygens and bearing his name, though perhaps independently devised by Campani of Rome. Probably four out of five astronomical eyepieces belong to this class.

The Huygenian ocular accomplishes two useful results—first, it gives a wider sharp field than any single lens, and second it compensates the chromatic aberration, which otherwise must be removed by a composite lens. It usually consists of a plano-convex lens, convex side toward the objective, which is brought inside the objective focus and forms an image in the plane of a rear diaphragm, and a similar eye lens of shorter focus by which this image is examined.

Fig. 100 shows the course of the rays—*A* being the field lens, *B* the diaphragm and *C* the eye lens. Let *1*, *2*, be rays which are incident near the margin of *A*. Each, in passing through the lens, is dispersed, the blue being more refracted than the red. Both rays come to a general focus at *B*, and, crossing, diverge slightly towards *C*.

But, on reaching *C*, ray *1*, that was nearer the margin and the more refracted because in a zone of greater pitch, now falls on *C* the nearer its center, and is less refracted than ray *2* which strikes *C* nearer the rim. If the curvatures of *A* and *C* are properly related *1* and *2* emerge from *C* parallel to each other and thus unite in forming a distinct image.

Now follow through the two branches of *1* marked 1_r, and 1_v, the red and violet components. Ray 1_v, the more refrangible, strikes C nearer the center, and is the less refracted, emerging from C substantially parallel with its mate 1_r, hence blending the red and violet images, if, being of the same glass, A and C have suitably related focal lengths and separation.

As a matter of fact the condition for this chromatic compensation is

$$d = \frac{f + f'}{2}$$

where d is the distance between the lenses and f, f′, their respective focal lengths. If this condition of achromatism be combined with that of equal refraction at A and C, favorable to minimizing the spherical aberration, we find $f = 3f'$ and $d = 2f'$. This is the conventional Huygenian ocular with an eye lens ⅓ the focus of the field lens, spaced at double the focus of the eye lens, with the diaphragm midway.

In practice the ratio of foci varies from 1:3 to 1:2 or even 1:1.5, the exact figure varying with the amount of over-correction in the objective and under-correction in the eye that has to be dealt with, while the value of d should be adjusted by actual trial on the telescope to obtain the best color correction practicable. One cannot use any chance ocular and expect the finest results.

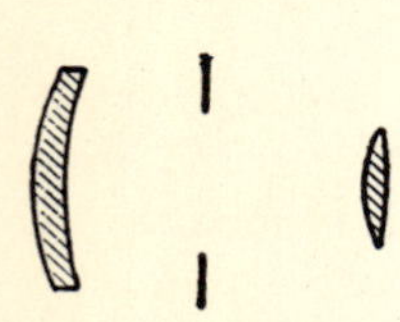

Fig. 101*a*.—Airy Ocular.

The Huygenian eyepieces are often referred to as "negative" inasmuch as they cannot be used directly as magnifiers, although dealing effectively with an image rather than an object. The statement is also often made that they cannot be used with cross wires. This is incorrect, for while there is noticeable distortion toward the edge of the wide field, to say nothing of astigmatism, in and near the center of the field the situation is a good deal better.

Central cross wires in the plane of the diaphragm are entirely suitable for alignment of the instrument, and over a moderate extent of field the distortion is so small that a micrometer scale in the plane of the diaphragm gives very good approximate measurements, and indeed is widely used in microscopy.

It should be noted that the achromatism of this type of eyepiece is compensatory rather than real. One cannot at the same time bring the images of various colors to the same size, and also to the same plane. As failure in the latter respect is comparatively unimportant, the Huygenian eyepiece is adjusted so far to compensate the paths of the various rays as to bring the colored images to the same size, and in point of fact the result is very good.

The field of the conventional form of Huygenian ocular is fully 40°, and the definition, particularly centrally, is very excellent. There are no perceptible ghosts produced, and while some 10% of light is lost by reflection in the extra lens it is diffused in the general field and is damaging only as it injures the contrast of faint objects. The theory of the Huygenian eyepiece was elaborately given by Littrow, (Memoirs R. A. S. Vol. 4, p. 599), wherein the somewhat intricate geometry of the situation is fully discussed.

Various modifications of the Huygenian type have been devised and used. Figure 101*a* is the Airy form devised as a result of a somewhat full mathematical investigation by Sir George Airy, later Astronomer Royal. Its peculiarity lies in the form of the lenses which preserve the usual 3:1 ratio of focal lengths. The field lens is a positive meniscus with a noticeable amount of concavity in the rear face while the eye lens is a "crossed" lens, the outer curvature being about ⅙ of the inner curvature. The marginal field in this ocular is a little better than in the conventional Huygenian.

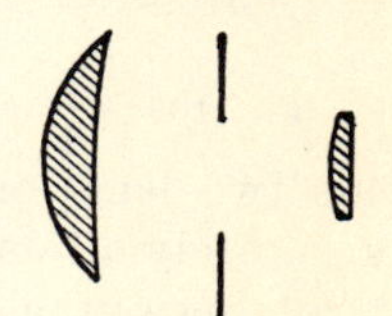

FIG. 101*b*.—Mittenzwey Ocular.

A commoner modification now-a-days is the Mittenzwey form, Fig. 101*b*. This is usually made with 2:1 ratio of focal lengths, and the field lens still a meniscus, but less conspicuously concave than in the Airy form. The eye lens is the usual plano-convex. It is widely used, especially abroad, and gives perhaps as large available field as any ocular yet devised, approximately 50°, with pretty good definition out to the margin.

Finally, we come to the solid eyepiece Fig. 102*a*, devised by the late R. B. Tolles nearly three quarters of a century ago, and and often made by him both for telescopes and microscopes. It is practically a Huygenian eyepiece made out of a single cylinder of glass with a curvature ratio of 1½:1 between the eye and the field lens. A groove is cut around the long lens at about ⅓ its length from the vertex of the field end. This serves as a stop, reducing the diameter of the lens to about one-half its focal length.

It is in fact a Huygenian eyepiece free from the loss of light in the usual construction. It gives a wide field, more extensive than in the ordinary form, with exquisite definition. It is really a most admirable form of eyepiece which should be used far more than is now the case. The late Dr. Brashear is on record as believing that all negative eyepieces less than ¾ inch focus should be made in this form.

So far as the writer can ascertain the only reason that it is not more used is that it is somewhat more difficult to construct than

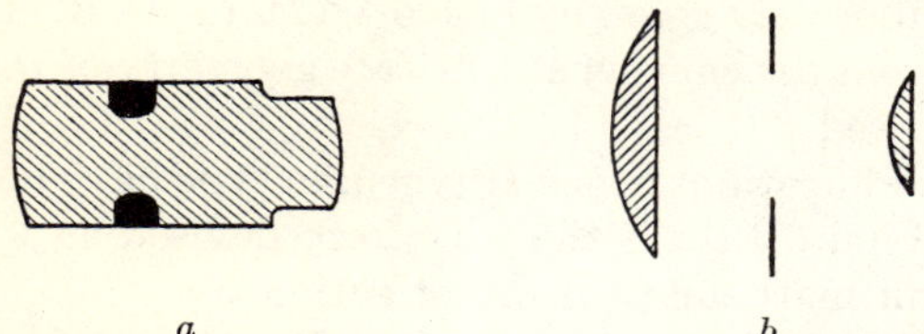

Fig. 102*a*.—Tolles Solid Ocular. Fig. 102*b*.—Compensated Ocular.

the two-lens form, for its curvatures and length must be very accurately adjusted. It is consequently unpopular with the constructing optician in spite of its conspicuous merits. It gives no ghosts, and the faint reflection at the eye end is widely spread so that if the exterior of the cylinder is well blackened, as it should be, it gives exceptional freedom from stray light. Still another variety of the Huygenian ocular sometimes useful is analogous to the compensating eyepiece used in microscopy. If, as commonly is the case, a telescope objective is over-corrected for color to correct for the chromatism of the eye in low powers, the high powers show strong over correction, the blue focus being longer than the red, and the blue image therefore the larger.

If now the field lens of the ocular be made of heavy flint glass and the separation of the lenses suitably adjusted, the stronger refraction of the field lens for the blue pulls up the blue focus and brings its image to substantially the dimensions of the red, so that the eye lens performs as if there were no overcorrection of the objective.

The writer has experimented with an ocular of this sort as shown in Fig. 102*b* and finds that the color correction is, as might be expected, greatly improved over a Mittenzwey ocular of the same focus (⅕ inch). There would be material advantage in thus varying the ocular color correction to suit the power.

In the Huyghenian eyepiece the equivalent focal length F is given by,

$$F = \frac{2\,ff'}{f + f'}$$

where f and f′ are the focal lengths of the field and eye lenses respectively. This assumes the normal spacing, d, of half the sum of the focal lengths, not always adhered to by constructors. The perfectly general case, as for any two combined lenses is,

$$F = \frac{ff_1}{f + f_1 - d}$$

To obtain a flatter field, and particularly one free from distortion the construction devised by Ramsden is commonly used.

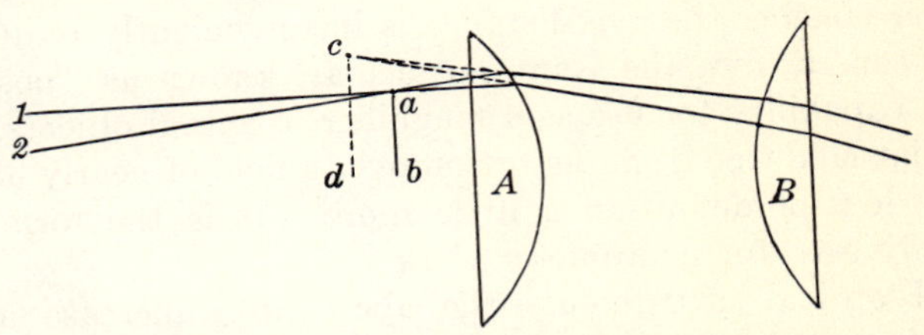

Fig. 103.—Path of Rays Through Ramsden Ocular.

This consists, Fig. 103, of two plano convex lenses of equal focal length, placed with their plane faces outward, at a distance equal to, or somewhat less than, their common focal length. The former spacing is the one which gives the best achromatic compensation since as before the condition for achromatism is

$$d = \tfrac{1}{2}(f + f')$$

When thus spaced the plane surface of the field lens is exactly in the focus of the eye lens, the combined focus F is the same as that of either lens, since as just shown in any additive combination of two lenses

$$F = \frac{ff'}{f + f' - d}$$

and while the field is flat and colorless, every speck of dust on the field lens is offensively in view.

It is therefore usual to make this ocular in the form suggested by Airy, in which something of the achromatic correction is sacrificed to obviate this difficulty, and to obtain a better balance of the residual aberrations. The path of the rays is shown in Fig. 103. The lenses *A* and *B* are of the same focal length but are now spaced at $\frac{2}{3}$ of this length apart.

The two neighboring rays *1*, *2*, coming through the objective from the distant object meet at the objective focus in a point, *a*, of the image plane *a b*. Thence, diverging, they are so refracted by *A* and *B* as to leave the latter substantially parallel so that both appear to proceed from the point c, of the image plane *c*, *d*, in the principal focus of *B*.

From the ordinary equation for the combination, $F = \frac{3}{4}$ f. The combination focusses $\frac{1}{4}$ f back of the principal focus of the objective, and the position of the eye is $\frac{1}{4}$ F back of the eye lens, which is another reason for shortening the lens spacing. At longer spacing the eye distance is inconveniently reduced.

Thus constituted, the Ramsden ocular, known as "positive" from its capability for use as a magnifier of actual objects, gives a good flat field free from distortion over a field of nearly 35° and at some loss of definition a little more. It is the form most commonly used for micrometer work.

In all optical instruments the aberrations increase as one departs from the axis, so that angular field is rather a loose term depending on the maximum aberrations that can be tolerated.[1]

Of the Ramsden ocular there are many modifications. Sometimes f and f′ are made unequal or there is departure from the simple plano-convex form. More often the lenses are made achromatic, thus getting rid of the very perceptible color in the simpler form and materially helping the definition. Figure 104*a* shows such an achromatic ocular as made by Steinheil. The general arrangement is as in the ordinary Ramsden, but the sharp field is slightly enlarged, a good 36°, and the definition is improved quite noticeably.

A somewhat analogous form, but considerably modified in

[1] The angular field a is defined by

$$\tan \tfrac{1}{2} a = \frac{\gamma}{F}$$

where γ is, numerically, the radius of the field sharp enough for the purpose in hand, and F the effective focal length of the ocular.

detail, is the Kellner ocular, Fig. 104*b*. It was devised by an optician of that name, of Wetzlar, who exploited it some three quarters of a century since in a little brochure entitled "Das orthoskopische Okular," as notable a blast of "hot air" as ever came from a modern publicity agent.

As made today the Kellner ocular consists of a field lens which is commonly plano-convex, plano side out, but sometimes crossed or even equiconvex, combined with a considerably smaller eye lens which is an over-corrected achromatic. The focal length of the field lens is approximately 7/4 F, that of the eye lens 4/3 F, separated by about 3/4 F.

This ocular has its front focal plane very near the field lens, sometimes even within its substance, and a rather short eye distance, but it gives admirable definition and a usable field of very great extent, colorless and orthoscopic to the edge. The

Fig. 104.—Achromatic and Kellner Oculars.

writer has one of 2⅝″ focus, with an achromatic triplet as eye lens, which gives an admirable field of quite 50°.

The Kellner is decidedly valuable as a wide field positive ocular, but it has in common with the two just previously described a sometimes unpleasant ghost of bright objects. This arises from light reflected from the inner surface of the field lens, and back again by the front surface to a focus. This focus commonly lies not far back of the field lens and quite too near to the focus of the eye lens for comfort. It should be watched for in going after faint objects with oculars of the types noted.

A decidedly better form of positive ocular is the modern orthoscopic as made by Steinheil and Zeiss, Fig. 105*a*. It consists of a triple achromatic field lens, a dense flint between two crowns, with a plano-convex eye lens of much shorter focus (⅓ to ½) almost in contact on its convex side.

The field triplet is heavily over-corrected for color, the front focal plane is nearly ½ F ahead of the front vertex of the field lens, and the eye distance is notably greater than in the Kellner. The field is above 40°, beautifully flat, sharp, and orthoscopic, free of troublesome ghosts. On the whole the writer is inclined to rate it as the best of two-lens oculars.

There should also here be mentioned a very useful long relief ocular, often used for artillery sights, and shown in Fig. 105*b*. It consists like Fig. 104*a*, of a pair of achromatic lenses, but they are placed with the crowns almost in contact and are frequently used with a simple plano covex field lens of much longer focus, to render the combination more fully orthoscopic.

The field, especially with the field lens, is wide, quite 40° as apparent angle for the whole instrument, and the eye distance is roughly equal to the focal length. It is a form of ocular that

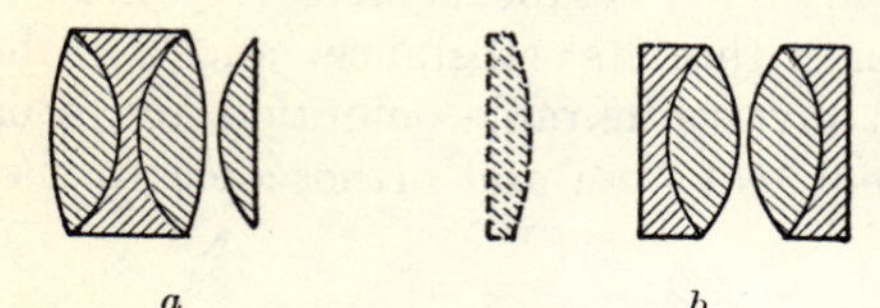

FIG. 105.—Orthoscopic and Long Relief Oculars.

might be very advantageously used in finders, where one often has to assume uncomfortable angles of view, and long relief is valuable.

Whatever the apparent angular field of an ocular may be, the real angular field of view is obtained by dividing the apparent field by the magnifying power. Thus the author's big Kellner, just mentioned, gives a power of 20 with the objective for which it was designed, hence a real field of 2½°, while a second, power 65, gives a real field of hardly 0°40′, the apparent field in this case being a trifle over 40°. There is no escaping this relation, so that high power always implies small field.

The limit of apparent field is due to increasing errors away from the axis, strong curvature of the field, and particularly astigmatism in the outer zones. The eye itself can take in only about 40° so that more than this, while attainable, can only be utilized by peering around the marginal field.

For low powers the usable field is helped out by the accommodation of the eye, but in oculars of short focus the curvature of field is the limiting factor. The radius of curvature of the image is, in a single lens approximately $\frac{3}{2}$ F, and in the common two-lens forms about $\frac{3}{4}$ F.

In considering this matter Conrady has shown (M. N. **78** 445) that for a total field of 40° the sharpness of field fails at a focal length of about 1 inch for normal power of accommodation. The best achromatic combinations reduce this limit to about ½ inch.

At focal lengths below this the sharpest field is obtainable only with reduced aperture. There is an interesting possibility of building an anastigmatic ocular on the lines of the modern photographic lens, which Conrady suggests, but the need of wide field in high powers is hardly pressing enough to stimulate research.

Finally we may pass to the very simple adjunct of most small telescopes, the terrestrial ocular which inverts the image and shows the landscape right side up. Whatever its exact

Fig. 106.—Ordinary Terrestrial Ocular.

form it consists of an inverting system which erects the inverted image produced by the objective alone, and an eyepiece for viewing this erected image. In its common form it is composed of four plano-convex lenses arranged as in Fig. 106. Here A and B form the inverting pair and C and D a modified Huygenian ocular. The image from the objective is formed in the front focus of AB which is practically an inverted ocular, and the erected image is formed in the usual way between C and D.

The apparent field is fairly good, about 35°, and while slightly better corrections can be gained by using lenses of specially adjusted curvatures, as Airy has shown, these are seldom applied. The chief objection to this erecting system is its length, some ten times its equivalent focus. Now and then to save light and gain field, the erector is a single cemented combination and the ocular like Fig. 99*a* or Fig. 102*a*. Fig. 107 shows a terrestrial eyepiece so arranged, from an example by the late R. B. Tolles. When carefully designed an apparent field of 40° or more can be secured, with great brilliancy, and the length of the erecting system is moderate.

Very much akin in principle is the eyepiece microscope, such as is made by Zeiss to give variable power and a convenient position of the eye in connection with filar micrometers, Fig. 108.

It is provided with a focussing collar and its draw tube allows varying power just as in case of an ordinary microscope. In fact eyepiece microscopes have long been now and then used to advantage for high powers. They are easier on the eye, and give

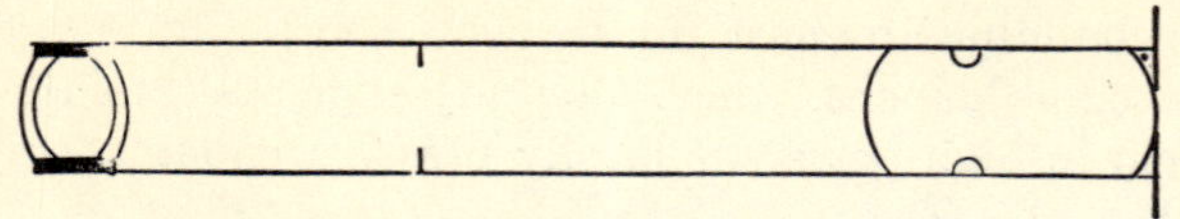

Fig. 107.—Tolles Triplet Inverting System.

greater eye distance than the exceedingly small eye lenses of short focus oculars, and using a solid eyepiece and single lens objective lose no more light than an ordinary Huygenian ocular. The erect resultant image is occasionally a convenience in astronomical use.

Quite analogous to the eyepiece microscope is the so-called

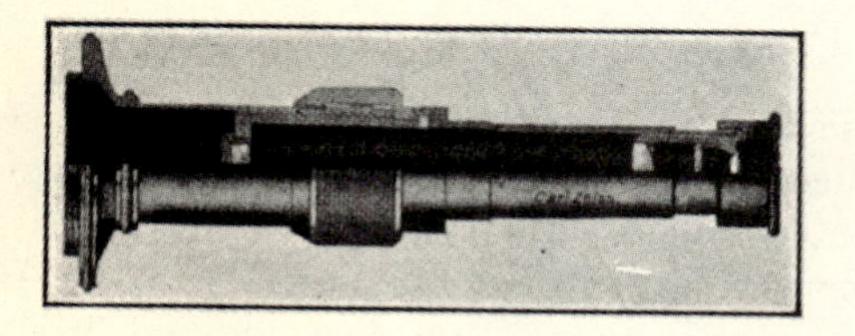

Fig. 108.—Microscope as Ocular.

"Davon" micro-telescope. Originally developed as an attachment for the substage of a microscope to give large images of objects at a little distance it has grown also into a separate

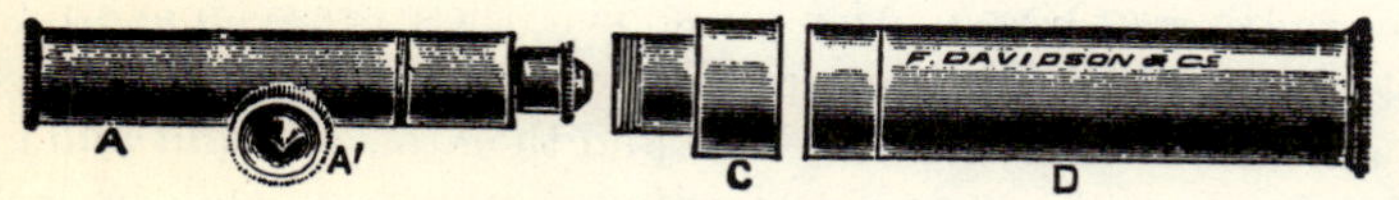

Fig. 109.—"Davon" Instrument.

hand telescope, monocular or binocular, for general purposes. The attachment thus developed is shown complete in Fig. 109. D is merely a well corrected objective set in a mount provided with ample stops. The image is viewed by an ordinary microscope or special eyepiece microscope A, as the case may be, furnished with rack focussing at A′ and assembled with the objective by means of the carefully centered coupling C.

It furnishes a compact and powerful instrument, very suitable for terrestrial or minor astronomical uses, like the Tolles' short-focus hand telescopes already mentioned. When properly designed telescopes of this sort give nearly the field of prism glasses, weigh much less and lose far less light for the same effective power and aperture. They also have under fairly high powers rather the advantage in the matter of definition, other things being equal.

CHAPTER VII

HAND TELESCOPES AND BINOCULARS

The hand telescope finds comparatively little use in observing celestial bodies. It is usually quite too small for any except very limited applications, and cannot be given sufficient power without being difficult to keep steady except by the aid of a fixed mounting. Still, for certain work, especially the observation of variable stars, it finds useful purpose if sufficiently compact and of good light-gathering power.

There is most decidedly a limit to the magnifying power which can be given to an instrument held in the hand without making the outfit too unsteady to be serviceable. Anything beyond 8 to 10 diameters is highly troublesome, and requires a rudimentary mount or at least steadying the hand against something in order to observe with comfort.

The longer the instrument the more difficult it is to manage, and the best results with hand telescopes are to be obtained with short instruments of relatively large diameter and low power. The ordinary field glass of Galilean type comes immediately to mind and in fact the field glass is and has been much used. As ordinarily constructed it is optically rather crude for astronomical purposes. The objectives are rarely well figured or accurately centered and a bright star usually appears as a wobbly flare rather than a point.

Furthermore the field is generally small, and of quite uneven illumination from centre to periphery, so that great caution has to be exercised in judging the brightness of a star, according to its position in the field. The lens diameter possible with a field glass of ordinary construction is limited by the limited distance between the eyes, which must be well centered on the eyepieces to obtain clear vision.

The inter-pupillary distance is generally a scant 2½ inches so that the clear aperture of one of the objectives of a field glass is rarely carried up to 2 inches. The best field glasses have each objective a triple cemented lens, and the concave lenses also triplets, the arrangement of parts being that shown in Fig. 110. Glasses of this sort rarely have a magnifying power above 5.

In selecting a field glass with the idea of using it on the sky try it on a bright star, real or artificial, and if the image with careful focussing does not pull down to a pretty small and uniform point take no further interest in the instrument.

The advantage of a binocular instrument is popularly much exaggerated. It gives a somewhat delusive appearance of brilliancy and clearness which is psychological rather than

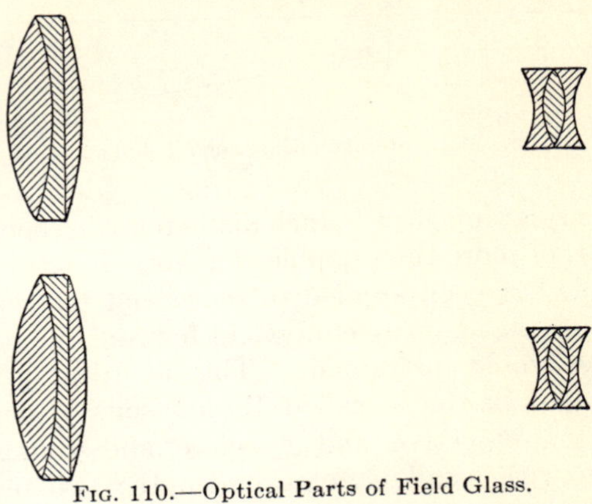

Fig. 110.—Optical Parts of Field Glass.

physical. During the late war a very careful research was made at the instance of the United States Government to determine the actual value of a binocular field glass against a monocular one of exactly the same type, the latter being cheaper, lighter, and in many respects much handier.

The difference found in point of actual seeing all sorts of objects under varying conditions of illumination was so small as to be practically negligible. An increase of less than 5 per cent in magnifying power enabled one to see with the monocular instrument everything that could be seen with the binocular, equally well, and it is altogether probable that in the matter of seeing fine detail the difference would be even less than in general use, since it is not altogether easy to get the two sides of a binocular working together efficiently or to keep them so afterwards.

There has been, therefore, a definite field for monocular hand telescopes of good quality and moderate power and such are manufactured by some of the best Continental makers. Such instruments have sometimes been shortened by building them on the exact principle of the telephoto lens, which gives a relatively large image with a short camera extension.

A much shortened telescope, as made by Steinheil for solar photographic purposes, is shown in Fig. 111. This instrument with a total length of about 2 feet and a clear aperture of $2\frac{3}{8}$

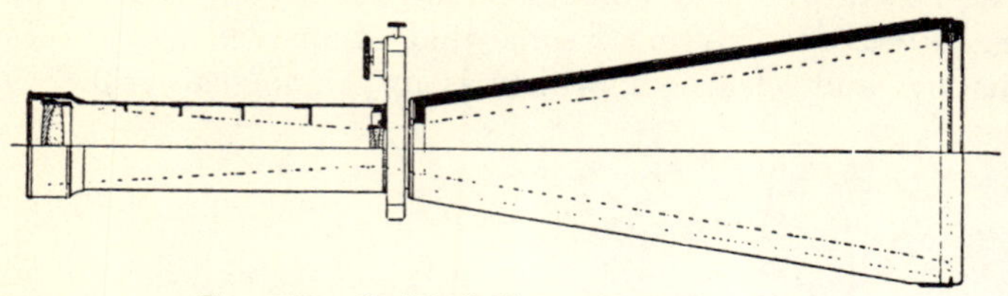

Fig. 111.—Steinheil Shortened Telescope.

inches gives a solar image of $\frac{1}{2}$ inch diameter, corresponding to an ordinary glass of more than double that total length. Quite the same principle has been applied to terrestrial telescopes by the same maker, giving again an equivalent focus of about double the length of the whole instrument. This identical principle has often been used in the so-called Barlow lens, a negative lens placed between objective and eyepiece and giving increased magnification with small increase of length; also photographic enlargers of substantially similar function have found considerable use.

A highly efficient hand telescope for astronomical purposes might be constructed along this line, the great shortening of the instrument making it possible to use somewhat higher powers than the ordinary without too much loss of steadiness. There is also constructed a binocular for strictly astronomical use consisting of a pair of small hand comet-seekers.

One of these little instruments is shown in Fig. 112. It has a clear diameter of objectives of $1\frac{3}{8}$ inch, magnification of 5, and a brilliant and even field of $7\frac{1}{2}$ degrees. The objectives are triplets like Fig. 57, already referred to, the oculars achromatic doublets of the type of Fig. 104*a*.

With the exception of these specialized astronomical field glasses the most useful and generally available hand instrument is the prism glass now in very general use. It is based on reversal of the image by internal total reflection in two prisms having their reflecting surfaces perpendicular each to the other. The rudiments of the process lie in the simple reversion prism shown in diagram in Fig. 113.

This is nothing more nor less than a right angled glass prism set with its hypothenuse face parallel and with its sides at 45° to the optical axis of the instrument. Rays falling upon one of its refracting faces at an angle of 45° are refracted upon the hypothenuse face, are there totally reflected and emerge from the second face of the prism parallel to their original course.

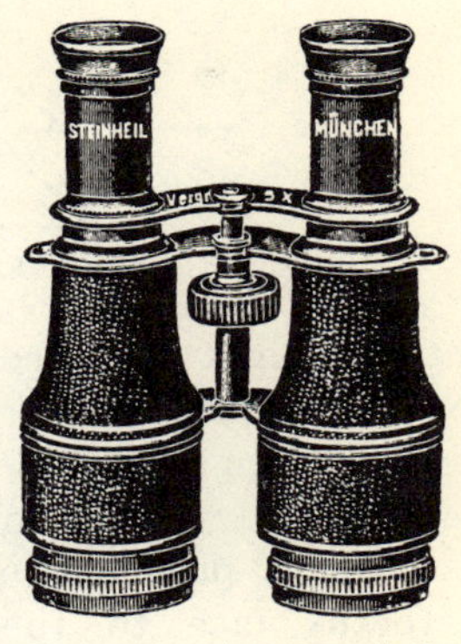

Fig. 112.—Astronomical Binocular.

Inspection of Fig. 113 shows that an element like A B perpendicular to the plane of the hypothenuse face is inverted by the total reflection so that it takes the position A′ B′. It is equally clear that an element exactly perpendicular to A B will be reflected from the hypothenuse face flat-wise as it were, and will emerge without its ends being reversed so that the net effect of this single reflection is to invert the image without reversing it laterally at the same time.

On the other hand if a second prism be placed behind the first, flat upon its side, with its hypothenuse face occupying a plane exactly perpendicular to that of the first prism, the line A′B′ will be refracted, totally reflected and refracted again out of the prism without a second inversion, while a line perpendicular to A′B′ will be refracted endwise on the hypothenuse face of the second prism and will be inverted as was the line A B at the start.

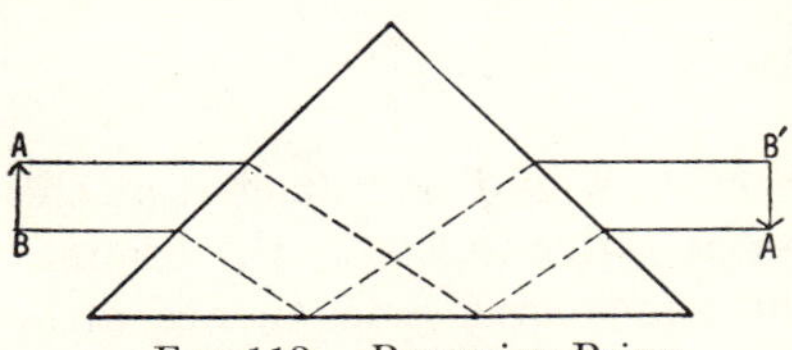

Fig. 113.—Reversion Prism.

Consequently two prisms thus placed will completely invert the image, producing exactly the same effect as the ordinary inverting system Fig. 106. The simple reversion prism is useful as furnishing a means, when placed over an eye lens, and rotated, of revolving the image on itself, a procedure occasionally convenient, especially in stellar photometry. The two prisms together constitute a true inverting system and have been utilized in that function, but they give a rather small angular field and have never come into a material amount of use. The exact effect of this combination, known historically as Dove's prisms, is shown plainly in Fig. 114.

The first actual prismatic inverting system was due to M. Porro, who invented it about the middle of the last century, and

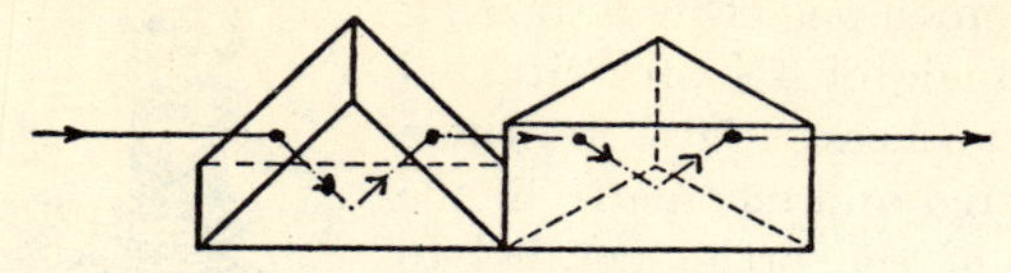

Fig. 114.—Dove's Prisms.

later brought it out commercially under the name of "Lunette à Napoléon Troisième," as a glass for military purposes.

The prism system of this striking form of instrument is shown in Fig. 115. It was composed of three right angle prisms *A*, *B*, and *C*. *A* presented a cathetus face to the objective and *B* a cathetus face to the ocular. Obviously a vertical element

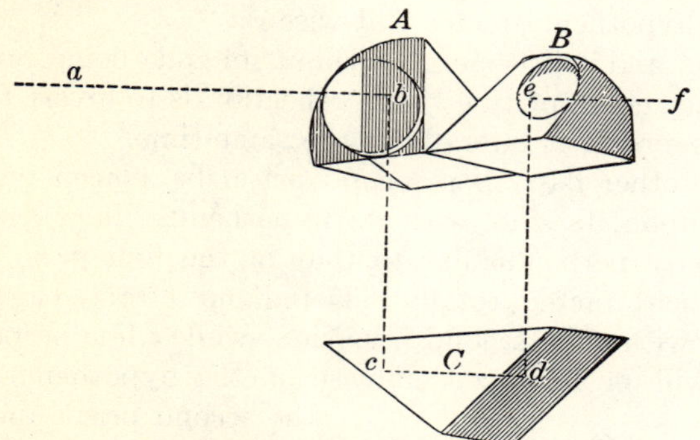

Fig. 115.—Porro's Prism System.

brought in along *a* from the objective would be reflected at the hypothenuse face *b*, to a position at right angles to the original one, would enter the hypothenuse face of *C* and thence after two reflections at *c* and *d* flatwise and without change of direction would emerge, enter the lower cathetus face of *B* and by reflection at the hypothenuse face *e* of *B* would be turned another 90° making a complete reversion as regards up and down at the eye placed at *f*. An element initially at right angles to the one just considered would enter *A*, be reflected flatwise, in the faces of *C* be twice reflected endwise, thereby completely inverting it, and would again be reflected flatwise from the hypothenuse face of C, thereby effecting, as the path of the rays indicated plainly shows, a complete inversion of the image. Focussing was very simply attained by a screw motion affecting the prism C and the whole affair was in a small flat case, the external appearance and size of which is indicated in Fig. 116.

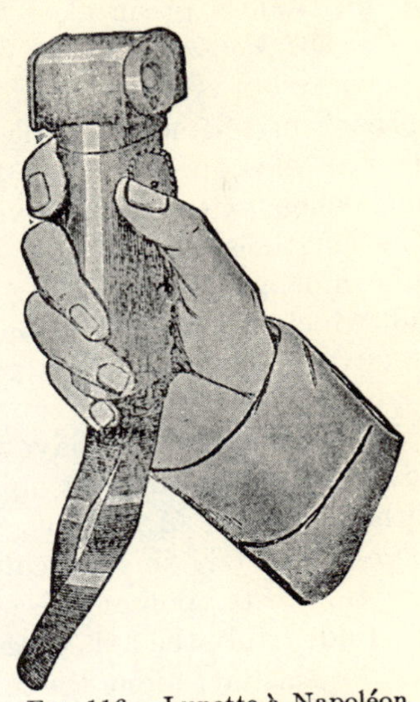

Fig. 116.—Lunette à Napoléon Troisième.

From ocular to objective the length was about an inch and a half. It was of 10 power and took in a field of 45 yards at a distance of 1000 yards. Here for the first time we find a prismatic inverting system of strictly modern type. And it is interesting to note that if one had wished to make a binocular "Lunette à Napoléon Troisième" he would inevitably have produced an instrument with enhanced stereoscopic effect like the modern prism field glass by the mere effort to dodge the observer's nose. Somewhat earlier M. Porro had arranged his prisms in the present conventional form of Fig. 117, where two right angle prisms have their faces positioned in parallel planes,

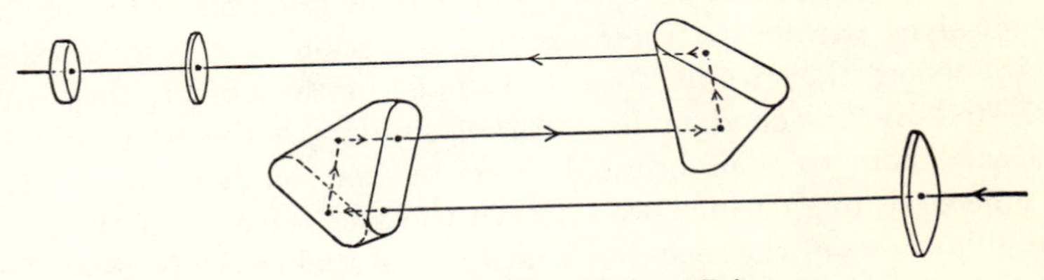

Fig. 117.—Porro's First Form of Prisms.

but turned around by 90° as in Fig. 114. The ray traced through this conventional system shows that exactly the same inversion occurs here as in the original Porro construction, and this form is the one which has been most commonly used for prismatic inversion and is conveniently known as Porro's first form, it actually having been antecedent in principle and practice to the

"Lunette à Napoléon Troisième." The original published description of Porro's work, translated from "Cosmos" Vol. 2, p. 222 (1852) et seq. is here annexed as it sets forth the origin of the modern prism glass in unmistakable terms.

Cosmos, Vol. 2, p. 222.—"We have wished for some time to make known to our readers the precious advantages of the "longue-vue cornet" or télémetre of M. Porro. Ordinary spyglasses or terrestrial telescopes of small dimensions are at least 30 or 40 c.m. long when extended to give distinct vision of distant objects. The length is considerably reduced by substituting for a fixed tube multiple tubes sliding into each other. But the drawing out which this substitution necessitates is a somewhat grave inconvenience; one cannot point the telescope without arranging it and losing time.

For a long time we have wished it were possible to have the power of viewing distant objects, with telescopes very short and without draw. M. Porro's "longue-vue cornet" seems to us to solve completely this difficult and important problem. Its construction rests upon an exceedingly ingenious artifice which literally folds triply the axis of the telescope and the luminous ray so that by this fact alone the length of the instrument is reduced by two-thirds.

Let us try to give an idea of this construction: Behind the telescope objective M. Porro places a rectangular isosceles prism of which the hypothenuse is perpendicular to the optic axis. The luminous rays from the object fall upon the rectangular faces of this prism, are twice totally reflected, and return upon themselves parallel to their original direction: half way to the point where they would form the image of the object, they are arrested by a second prism entirely similar to the first, which returns them to their original direction and sends them to the eyepiece through which we observe the real image. If the rectangular faces of the second prism were parallel to the faces of the first, this real image would be inverted—the telescope would be an astronomical and not a terrestrial telescope. But M. Porro being an optician eminently dextrous, well divined that to effect the reinversion it sufficed to place the rectangular faces of the second prism perpendicular to the corresponding faces of the first by turning them a quarter revolution upon themselves.

In effect, a quarter revolution of a reflecting surface is a half revolution for the image, and a half revolution of the image evidently carries the bottom to the top and the right to the left, effecting a complete inversion. As the image is thus *redressed* independently of the eyepiece one can evidently view it with a simple two-lens ocular which decreases still further the length of the telescope so that it is finally reduced to about a quarter of that of a telescope of equal magnifying power, field and clearness.

The new telescope is then a true pocket telescope even with a magnifying power of 10 or 15. Its dimensions in length and bulk are those of a field glass usually magnifying only 4 to 6 times. The more draws, the more bother,—it here suffices to turn a little thumbscrew to find in an instant the point of sharpest vision.

In brilliancy necessarily cut down a little, not by the double total reflection, which as is well known does not lose light, but by the quadruple passage across the substance of the two prisms, the cornet in sharpness and amplification of the images can compare with the best hunting telescopes of the celebrated optician Ploessl of Vienna. M. Porro has constructed upon the same principles a marine telescope only 15 c.m. long with an objective of 40 m.m. aperture which replaces an ordinary marine glass 70 c.m. long. He has done still better,—a telescope only 30 c.m. long carries a 60 m.m. objective and can be made by turns a day and a night glass, by substituting by a simple movement of the hand and without dismounting anything, one ocular for the other. Its brilliancy and magnification of a dozen times with the night ocular, of twenty-five times with the day ocular, permits observing without difficulty the eclipses of the satellites of Jupiter.

This is evidently immense progress. One of the most illustrious of German physicists, M. Dove of Berlin, gave in 1851 the name of reversion prism to the combination of two prisms placed normally one behind the other so that their corresponding faces were perpendicular. He presented this disposition as an important new discovery made by himself. He doubtless did not know that M. Porro, who deserves all the honor of this charming application, had realized it long before him."

A little later M. Porro produced what is commonly referred to as Porro's second form, which is derived directly from annexing *A* Fig. 115 to the corresponding half of *C* as a single prism, the other half of *B* being similarly annexed to the prism *C*, thus forming two sphenoid prisms, such as are shown in Fig. 118 which may be mounted separately or may have their faces cemented together to save loss of light by reflections. The sphenoid prisms have had the reputation of being much more difficult to construct than the plain right angled prisms of the other forms shown. In point of fact they are not particularly difficult to make and the best inverting eyepieces for telescopes are now constructed with sphenoid prisms like those just described.

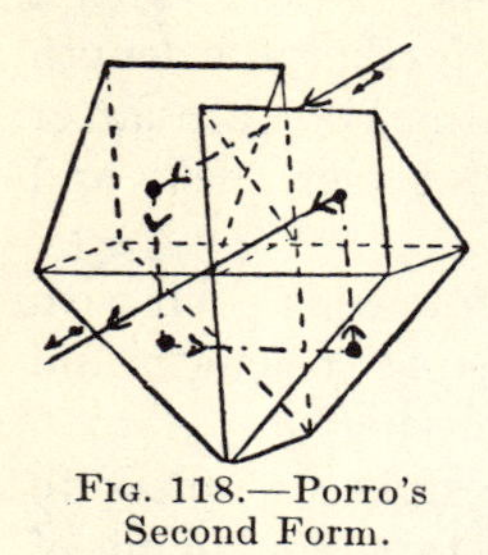

Fig. 118.—Porro's Second Form.

This particular arrangement lends itself very readily to a fairly compact and symmetrical mounting, as is well shown in Fig. 119

Fig. 119.—Clark Prismatic Eyepiece.

which is the terrestrial prismatic eyepiece as constructed by the Alvan Clark corporation for application to various astronomical telescopes of their manufacture. A glance at the cut shows the compactness of the arrangement, which actually shortens the linear distance between objective and ocular by the amount of the path of the ray through the prisms instead of lengthening the distance as in the common terrestrial eyepiece.

The field moreover is much larger than that attainable by a construction like Fig. 110, extending to something over 40°, and there is no strong tendency for the illumination or definition to fall off near the edge of the field.

In the practical construction of prism field glasses the two right angled prisms are usually separated by a moderate space as in Porro's original instruments so as to shorten the actual length

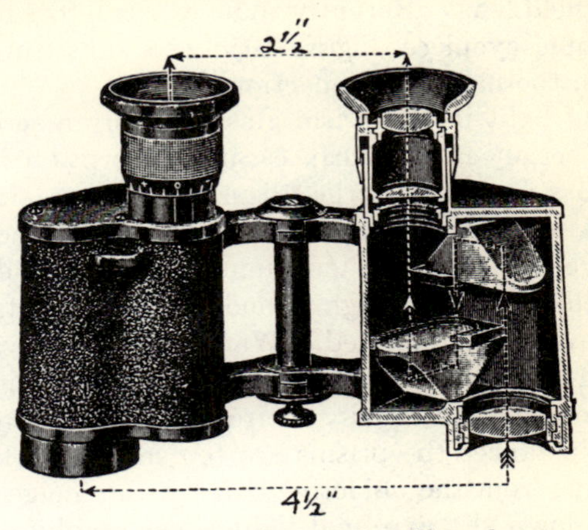

Fig. 120.—Section of Prism Binocular.

of the prism telescope by folding the ray upon itself as in Fig. 120, which is a typical modern binocular of this class.

The path of the rays is plainly shown and the manner in which the prisms fold up the total focal length of the objective is quite obvious. The added stereoscopic effect obtained by the arrangement of the two sides of the instrument is practically a very material gain. It gives admirable modelling of the visible field, a perception of distance which is at least very noticeable and a certain power of penetration, as through a mass of underbrush, which results from the objectives to a certain extent seeing around small objects so that one or the other of them gives an image of something beyond. For near objects there is some exaggeration of stereoscopic effect but on the whole for terrestrial use the net gain is decidedly in evidence.

A well made prism binocular is an extremely useful instrument for observation of the heavens, provided the objectives are of fair size, and the prisms big enough to receive the whole beam from the objective, and well executed enough to give a thoroughly good image with a flat field.

The weak points of the prism glass are great loss of light through reflection at the usual 10 air-glass surfaces and the general presence of annoying ghosts of bright objects in the field. Most such binoculars have Kellner eyepieces which are peculiarly bad, as we have seen, with respect to reflected images, and present the plane surface of the last prism to the plane front of the field lens. Recently some constructors have utilized the orthoscopic eyepiece, Figure 105*a*, as a substitute with great advantage in the matter of reflections.

The loss of light in the prism glass is really a serious matter, between reflection at the surfaces and absorption in the thick masses of glass necessary in the prisms. If of any size the transmitted light is not much over one-half of that received, very seldom above 60%. If the instrument is properly designed the apparent field is in the neighborhood of 45°, substantially flat and fairly evenly illuminated. Warning should here be given however that many binoculars are on the market in which the field is far from flat and equally far from being uniform.

In many instances the prisms are too small to take the whole bundle of rays from the objective back to the image plane without cutting down the marginal light considerably. Even when the field is apparently quite flat this fault of uneven illumination may exist, and in a glass for astronomical uses it is highly objectionable.

Before picking out a binocular for a study of the sky make very careful trial of the field with respect to flatness and clean definition of objects up to the very edge. Then judge as accurately as you may of the uniformity of illumination, if possible by observation on two stars about the radius of the field apart. It should be possible to observe them in any part of the field without detectable change in their apparent brilliancy.

If the objectives are easily removable unscrew one of them to obtain a clear idea as to the actual size of the prisms.[1] Look out, too, for ghosts of bright stars.

[1] There are binoculars on the market which are to outward appearance prism glasses, but which are really ordinary opera glasses mounted with intent to deceive, sometimes bearing a slight variation on the name of some well known maker.

The objectives of prism glasses usually run from ¾ inch to 1½ inch in diameter, and the powers from 6 to 12. The bigger the objectives the better, provided the prisms are of ample size, while higher power than 6 or 8 is generally unnecessary and disadvantageous. Occasional glasses of magnifying power 12 to 20 or more are to be found but such powers are inconveniently great for an instrument used without support. Do not forget that a first class monocular prism glass is extremely convenient and satisfactory in use, to say nothing of being considerably less in price than the instrument for two eyes. A monocular prism

Fig. 121.—Binocular with Extreme Stereoscopic Effect.

glass, by the way, makes an admirable finder when fitted with coarse cross lines in the eyepiece. It is particularly well suited to small telescopes without circles.

Numerous modifications of Porro's inverting prisms have been made adapting them to different specific purposes. Of these a single familiar example will suffice as showing the way in which the Porro prism system can be treated by mere rearrangement of the prismatic elements. In Fig. 121 is shown a special Zeiss binocular capable of extreme stereoscopic effect. It is formed of two Porro prism telescopes with the rays brought into the objectives at right angles to the axis of the instrument by a right angled prism external to the objective.

The apertures of these prisms appear pointing forward in the cut. As shown they are in a position of maximum stereoscopic effect.

Being hinged the tubes can be swung up from the horizontal position, in which latter the objectives are separated by something like eight times the interocular distance. The stereoscopic effect with the tubes horizontal is of course greatly exaggerated so that it enables one to form a fair judgment as to the relative position of somewhat distant objects, a feature useful in locating shell bursts.

The optical structure of one of the pair of telescopes is shown in Fig. 122 in which the course of the entering ray can be traced through the exterior prism of the objective and the remainder of the reversing train and thence through the eyepiece. This prism erecting system is obviously derived from the "Lunette à Napoléon Troisième" by bringing down the prism *B* upon the corresponding half of *C* (Fig. 115) and cementing it thereto, meanwhile placing the objective immediately under *A*.

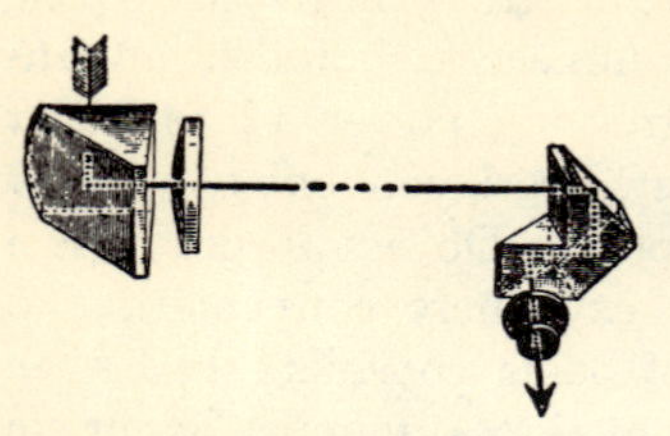

FIG. 122.—Path of Ray in Fig. 121.

One occasionally meets prismatic inverting systems differing considerably from the Porro forms. Perhaps the best known of these is the so called roof prism due to Prof. Abbé, Fig. 123, and occasionally useful in that the entering and emerging rays lie in the same straight line, thus forming a direct vision system. Looking at it as we did at the Porro system a vertical element in front of the prism is reversed in reflection from the two surfaces a and b, while a corresponding horizontal element is reflected flatwise so far as these are concerned, but is turned end for end by reflection at the roof surfaces c and d, thus giving complete inversion.

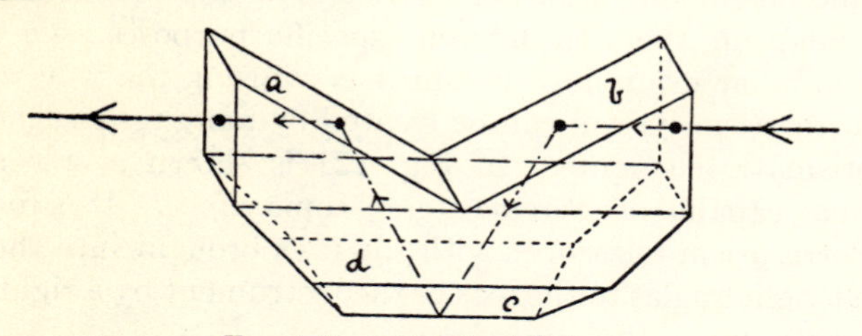

FIG. 123.—Abbé Roof Prism.

In practice the prism is made as shown, in three parts, two of them right angled prisms, the third containing the roof surfaces. The extreme precision required in figuring the roof forms a considerable obstacle to the production of such prisms in quantity and while they have found convenient use in certain special instruments like gunsights, where direct vision is useful, they are not extensively employed for general purposes, although both monocular and binocular instruments have been constructed by their aid.

One other variety of prism involving the roof principle has found some application in field glasses manufactured by the firm of Hensoldt. The prism form used is shown in Fig. 124. This like other forms of roof prism is less easy to make than the conventional Porro type. Numerous inverting and laterally reflecting prisms are in use for specific purposes. Some of them are highly ingenious and remarkably well adapted for their use, but hardly can be said to form a material portion of telescope practice. They belong rather to the technique of special instruments like gunsights and periscopes, while some of them have been devised chiefly as ingenious substitutes for the simpler Porro forms.

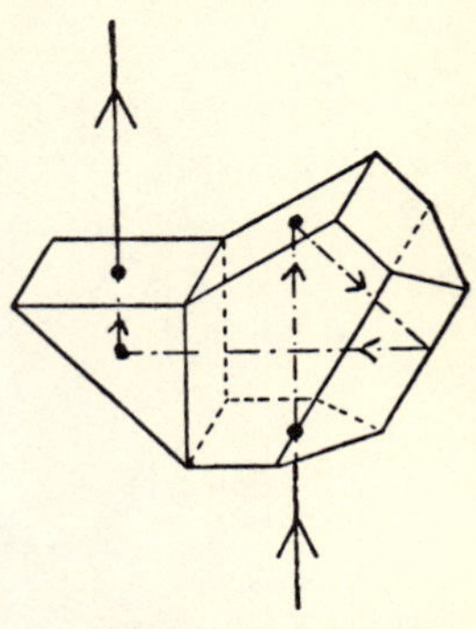

Fig. 124.—Hensoldt Prism.

Most prism telescopes both monocular and binocular are generally made on one or the other of the Porro forms. This is particularly true of the large binoculars which are occasionally constructed. Porro's second form with the sphenoid prisms seems to be best adapted to cases where shortening of the instrument is not a paramount consideration. For example, some Zeiss short focus telescopes are regularly made in binocular form, and supplied with inverting systems composed of two sphenoid prisms, and with oculars constructed on the exact principle of the triple nose-piece of a microscope, so that three powers are immediately available to the observer.

Still less commonly binocular telescopes of considerable aperture are constructed, primarily for astronomical use, being provided with prismatic inversion for terrestrial employment, but more particularly in order to gain by the lateral displacement of a Porro system the space necessary for two objectives of considerable size. As we have already seen, the practical diameter of objectives in a binocular is limited to a trifle over 2 inches unless space is so gained. The largest prismatic binocular as yet constructed is one made years ago by the Clarks, of 6¼ inches objective aperture and 92¼ inches focal length. So big and powerful an instrument obviously would give admirable binocular views of the heavens and so accurately was it constructed that the reports of its performance were exceedingly good. The same firm has made a good many similar binoculars of 3 inches and above, of which a typical example of 4-inch aperture and 60 inches focal length is shown in Fig. 125. In this case the erecting systems were of Porro's first form, and were provided with Kellner

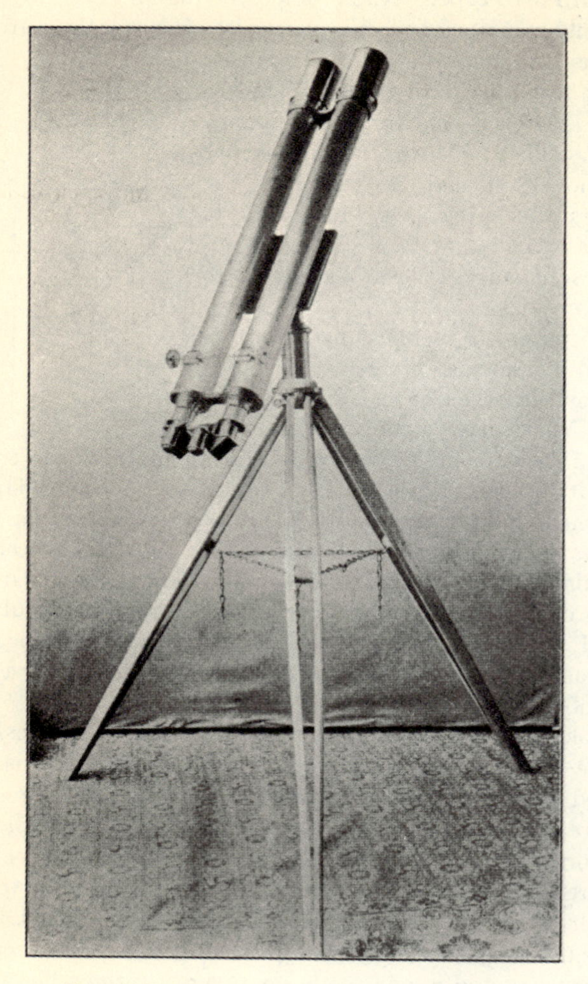

FIG. 125.—Clark 4-inch Binocular Telescope.

oculars of very wide field. These binoculars constructions in instruments of such size, however well made and agreeable for terrestrial observation, hardly justify the expense for purely astronomical use.

CHAPTER VIII

ACCESSORIES

Aside from the ordinary equipment of oculars various accessories form an important part of the observer's equipment, their number and character depending on the instrument in use and the purposes to which it is devoted.

First in general usefulness are several special forms of eyepiece equipment supplementary to the usual oculars. At the head of

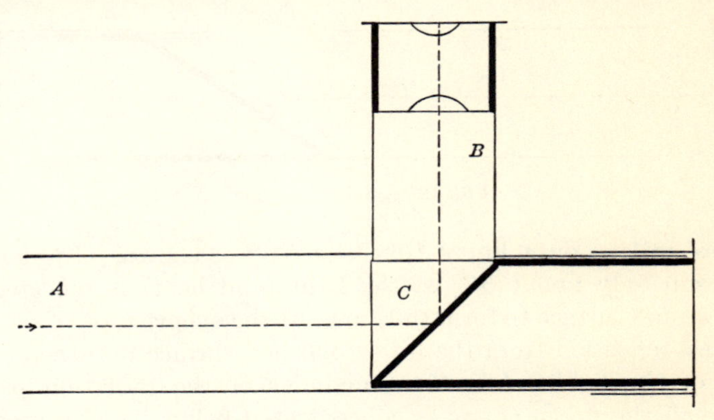

Fig. 126.—Star Diagonal.

the list is the ordinary star diagonal for the easier viewing of objects near the zenith here shown in Fig. 126. It is merely a tube, *A*, fitting the draw tube of the telescope, with a slotted side tube *B*, at a right angle, into which the ordinary ocular fits, and a right angled prism *C* with its two faces perpendicular respectively to the axes of the main and side tubes, and the hypothenuse face at 45° to each. The beam coming down the tube is totally reflected at this face and brought to focus at the ocular. The lower end of the tube is closed by a cap to exclude dust.

One looks, by help of this, horizontally at zenith stars, or, if observing objects at rather high altitude, views them at a comfortable angle downward. The prism must be very accurately made to avoid injury to the definition, but loses only about 10% of the light, and adds greatly to the comfort of observing.

Of almost equal importance is the solar diagonal devised by Sir John Herschel, Fig. 127. Here the tube structure *A*, *B*, is quite the same as in Fig. 126 but the right angled prism is replaced by a simple elliptical prism *C* of small angle, 10° or less, with its upper face accurately plane and at 45° to the axes of the

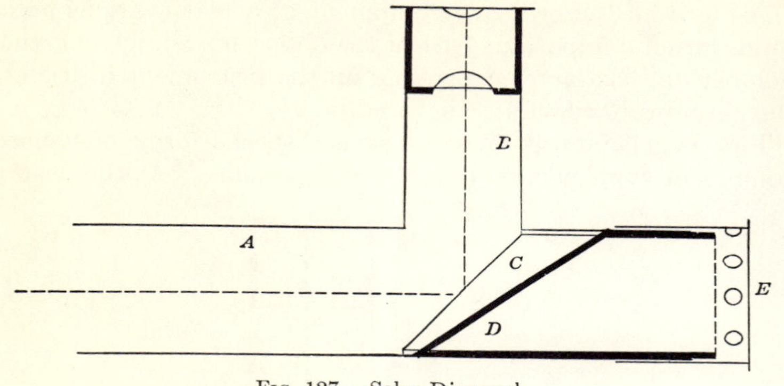

Fig. 127.—Solar Diagonal.

tubes, resting on a lining tube *D* cut off as shown. In viewing the sun only about 5% of the light (and heat) is reflected at this upper surface to form the image at the eyepiece.

Any reflection from the lower polished surface is turned aside out of the field, while the remainder of the radiation passes through the prism *C* and is concentrated below it. To prevent scorching the observer the lower end of the tube is capped at *E*, but the cap has side perforations to provide circulation for the heated air. Using such a prism, the remnant of light reflected can be readily toned down by a neutral tinted glass over the ocular.

In the telescopes of 3 inches and less aperture, and ordinary focal ratio, a plane parallel disc of very dark glass over the ocular gives sufficient protection to the eye. This glass is preferably of neutral tint, and commonly is a scant 1/16 inch thick. Some observers prefer other tints than neutral. A green and a red glass superimposed give good results and so does a disc of the deepest shade of the so-called Noviweld glass, which is similar in effect.

With an aperture as large as 3 inches a pair of superimposed dark glasses is worth while, for the two will not break simultaneously from the heat and there will be time to get the eye away in safety. A broken sunshade is likely to cost the observer a permanent scotoma, blindness in a small area of the retina which will neither get better nor worse as time goes on.

Above 3 inches aperture the solar prism should be used or, if one cares to go to fully double the cost, there is nothing more comfortable to employ in solar observation than the polarizing eyepiece, Fig. 128. This shows schematically the arrangement of the device. It depends on the fact that a ray of light falling on a surface of common glass at an angle of incidence of approximately 57° is polarized by the reflection so that while it is freely reflected if it falls again on a surface parallel to the first, it is absorbed if it falls at the same incidence on a surface at right angles to the first.

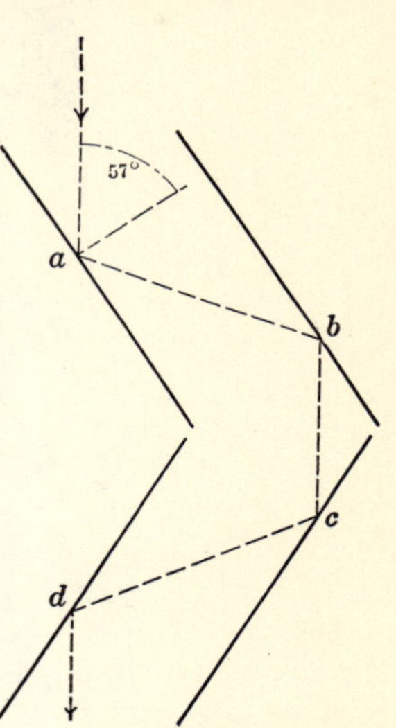

Fig. 128.—Diagram of Polarizing Eyepiece.

Thus in Fig. 128 the incident beam from the telescope falls on the black glass surface *a* at 57° incidence, is again reflected from the parallel mirror *b*, and then passed on, parallel to its original path, to the lower pair of mirrors *c*, *d*. The purpose of the second reflection is to polarize the residual light which through the convergence of the rays was incompletely polarized at the first.

The lower pair of mirrors *c*, *d*, again twice reflect the light at the polarizing angle, and, in the position shown, pass it on to the ocular diminished only by the four reflections. But if the second pair of mirrors be rotated together about a line parallel to *b c* as an axis the transmitted light begins to fade out, and when they have been turned 90°, so that their planes are inclined 90° to *a* and *b* (= 33° to the plane of the paper), the light is substantially extinguished.

Thus by merely turning the second pair of mirrors the solar image can be reduced in brilliancy to any extent whatever, without modifying its color in any way. The typical form given to the polarizing eyepiece is similar to Fig. 129. Here t_2 is the box containing the polarizing mirrors, a b, fitted to the draw tube, but for obvious reasons eccentric with it, t_1 is the rotating box containing the "analysing" mirrors c, d, and a is the ocular turning with it.

Sometimes the polarizing mirrors are actually a pair of Herschel prisms as in Fig. 127, facing each other, thus getting rid of much

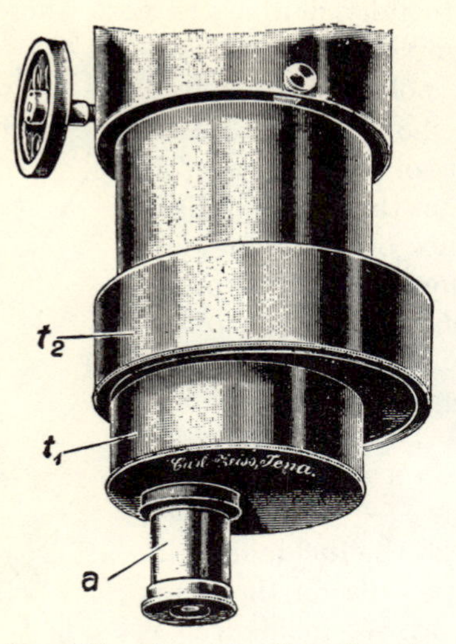

FIG. 129.—Polarizing Solar Eyepiece.

of the heat. Otherwise the whole set of mirrors is of black glass to avoid back reflections. In simpler constructions single mirrors are used as polarizer and analyser, and in fact there are many variations on the polarizing solar eyepiece involving about the same principles.

In any solar eyepiece a set of small diaphragms with holes from perhaps 1/64 inch up are useful in cutting down the general glare from the surface outside of that under scrutiny. These may be dropped upon the regular diaphragm of the ocular or conveniently arranged in a revolving diaphragm like that used with the older photographic lenses.

The measurement of celestial objects has developed a large group of important auxiliaries in the micrometers of very varied forms. The simplest needs little description, since it consists merely of a plane parallel disc of glass fitting in the focus of a positive ocular, and etched with a network of uniform squares, forming a reticulated micrometer by which the distance of one object from another can be estimated.

It can be readily calibrated by measuring a known distance or noting the time required for an equatorial star to drift across the squares parallel to one set of lines. It gives merely a useful approximation, and accurate measures must be turned over to more precise instruments.

The ring micrometer due, like so much other valuable apparatus, to Fraunhofer, is convenient and widely used for determining positions. It consists, as shown in Fig. 130, of an accurately

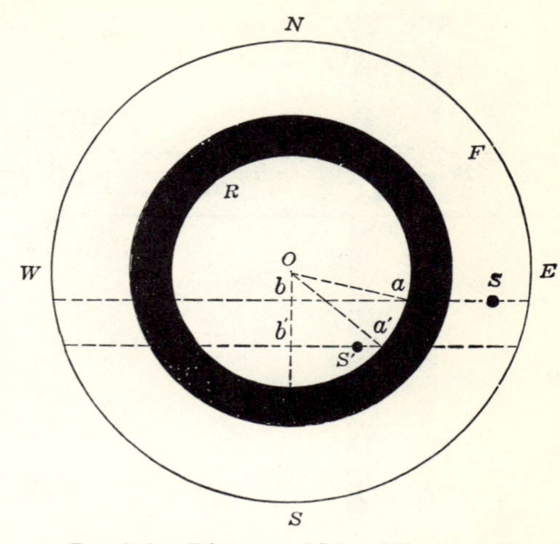

FIG. 130.—Diagram of Ring Micrometer.

turned opaque ring, generally of thin steel, cemented to a plane parallel glass or otherwise suspended in the center of the eyepiece field. The whole ring is generally half to two thirds the width of the field and has a moderate radial width so that both the ingress and the egress of a star can be conveniently timed.

It depends wholly on the measurement of time as the stars to be compared drift across the ring while the telescope is fixed, and while a clock or chronometer operating a sounder is a desirable adjunct one can do pretty well with a couple of stop watches since only differential times are required.

For full directions as to its use consult Loomis' Practical Astronomy, a book which should be in the library of every one who has the least interest in celestial observations. Suffice it to say here that the ring micrometer is very simple in use, and the computation of the results is quite easy. In Fig. 130 F is the edge of the field, R the ring, and *a b*, *a′b′*, the paths of the stars *s* and *s′*, the former well into the field, the latter just within the ring. The necessary data comprise the times of the transits at the edges of the ring, and the radius of the ring in angular measure, whence the difference in R. A. or Dec, can be obtained.[1]

Difference of R. A. = $\frac{1}{2}$ (t′ + t) − $\frac{1}{2}$ (T′ + T) where (T′ + T) refers to the transits of the first star. To obtain differences of declination, one declination should be known at least approxi-

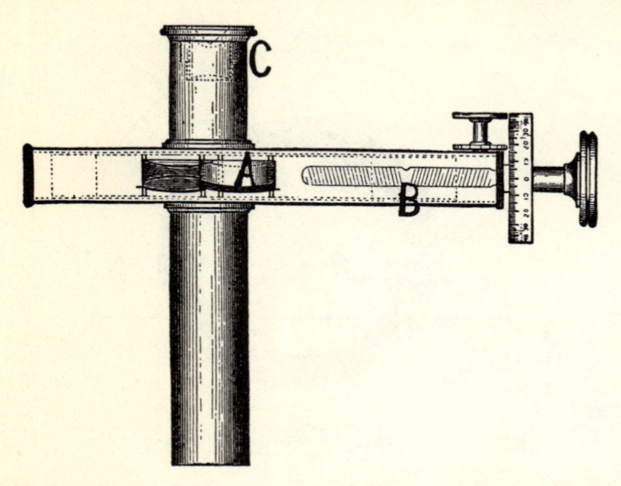

Fig. 131.—Double Image Micrometer. (*Courtesy of The Clarendon Press.*)

mately, and the second estimated from its relative position in the ring or otherwise. Then with these tentative values proceed as follows.

Put x = angle *a o b* and *x′* = angle *a′o′b′*
Also let d = approximate declination of *s* and
d′ = approximate declination of *s′*

$$\text{Then } \sin x = \frac{15}{2r} \cos d \, (T' - T)$$

$$\sin x' = \frac{15}{2r} \cos d' \, (t' - t) \text{ and finally}$$

[1] r the radius of the ring, is given by, $r = \frac{15}{2} (t' - t) \cos \text{Dec.}$, t′ − t being the seconds taken for central transit across the ring.

Difference of Dec. = r (cos x′ − cos x), when both arcs are on the same side of center of ring. If on opposite sides, Diff. = r (cos x′ + cos x).

There is also now and then used a square bar micrometer, consisting of an opaque square set with a diagonal in the line of diurnal motion. It is used in much the same way as the ring, and the reductions are substantially the same. It has some points of convenience but is little used, probably on account of the great difficulty of accurate construction and the requirement, for advantageous use, that the telescope should be on a well adjusted equatorial stand.[1] The ring micrometer works reasonably well on any kind of steady mount, requires no illumination of the field and is in permanent working adjustment.

Still another type of micrometer capable of use without a clock-drive is the double image instrument. In its usual form it is based on the principle that if a lens is cut in two along a diameter and the halves are slightly displaced along the cut all objects will be seen double, each half of the lens forming its own set of images.

Conversely, if one choses two objects in the united field these can be brought together by sliding the halves of the lens as before, and the extent of the movement needed measures the distance between them. Any lens in the optical system can be thus used, from the objective to the eyepiece. Fig. 131 shows a very simple double image micrometer devised by Browning many years ago. Here the lens divided is a so-called Barlow lens, a weak achromatic negative lens sometimes used like a telephoto lens to lengthen the focus and hence vary the power of a telescope.

This lens is shown at A with the halves widely separated by the double threaded micrometer screw B, which carries them apart symmetrically. The ocular proper is shown at C.

Double image micrometers are now mainly of historical interest, and the principle survives chiefly in the heliometer, a telescope with the objective divided, and provided with sliding mechanism of the highest refinement. The special function of the heliometer is the direct micrometric measurement of stellar distances too great to be within the practicable range of a filar micrometer—distances for example up to 1½° or even more.

The observations with the heliometer are somewhat laborious

[1] (For full discussion of this instrument see Chandler, Mem. Amer. Acad. Arts & Sci. 1885, p. 158).

and demand rather intricate corrections, but are capable of great precision. (See Sir David Gill's article "Heliometer" in the Enc. Brit. 11th Ed.). At the present day celestial photography, with micrometric measurement of the resulting plates, has gone far in rendering needless visual measurements of distances above a very few minutes of arc, so that it is somewhat doubtful whether a large heliometer would again be constructed.

The astronomer's real arm of precision is the filar micrometer. This is shown in outline in Fig. 132, the ocular and the plate that

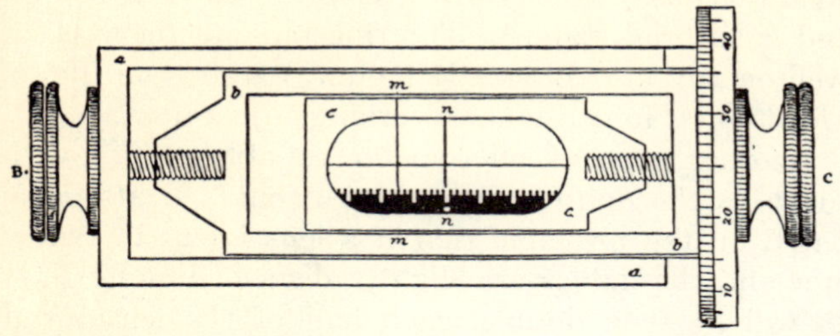

Fig. 132.—Filar Micrometer. (*Courtesy of J. B. Lippincott Co.*)

carries it being removed so as to display the working parts. It consists of a main frame aa, carrying a slide bb, which is moved by the screws and milled head B. The slide bb carries the vertical spider line mm, and usually one or more horizontal spider lines, line mm is the so-called fixed thread of the micrometer, movable only as a convenience to avoid shifting the telescope.

On bb moves the micrometer slide cc, carrying the movable spider line nn and the comb which records, with mm as reference line, the whole revolutions of the micrometer screw C. The ocular sometimes has a sliding motion of its own on cc, to get it positioned to the best advantage. In use one star is set upon mm by the screw B and then C is turned until nn bisects the other star.

Then the exact turns and fraction of a turn can be read off on the comb and divided head of C, and reduced to angular measure by the known constant of the micrometer, usually determined by the time of passage of a nearly equatorial star along the horizontal thread when mm, nn, are at a definite setting apart. (Then $r = \frac{15\,(t' - t)\cos d}{N}$ where r is the value of a revolution in seconds of arc, N the revolutions apart of mm, nn, and t and d as heretofore.)

Very generally the whole system of slides is fitted to a graduated circle, to which the fixed horizontal thread is diametral. Then by turning the micrometer until the horizontal threads cut the two objects under comparison, their position angle with reference to a graduted circle can be read off. This angle is conventionally counted from 0° to 360° from north around through east.

Figure 133 shows the micrometer constructed by the Clarks for their 24-inch equatorial of the Lowell Observatory. Here A is the head of the main micrometer screw of which the whole turns are

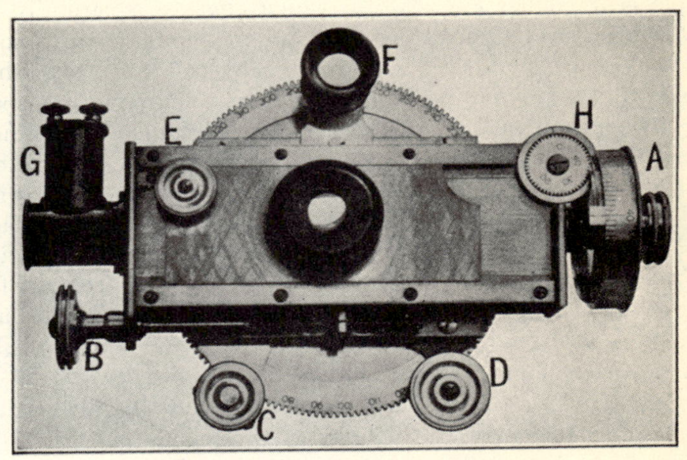

Fig. 133.—Filar Position Micrometer.

reckoned on the counter H in lieu of the comb of Fig. 132. B is the traversing screw for the fixed wire system, C the clamping screw of the position circle, D its setting pinion, E the rack motion for shifting the ocular, F the reading glass for the position circle, and G the little electric lamp for bright wire illumination. The parts correspond quite exactly with the diagram of Fig. 132 but the instrument is far more elegant in design than the earlier forms of micrometer and fortunately rid of the oil lamps that were once in general use. A small electric lamp with reflector throws a little light on the spider lines—just enough to show them distinctly. Or sometimes a faint light is thus diffused in the field against which the spider lines show dark.

Commonly either type of illumination can be used and modified as occasion requires. The filar micrometer is seldom used on small telescopes, since to work easily with it the instrument should be permanently mounted and clock-driven. Good work was done by some of the early observers without these aids, but at the cost of infinite pains and much loss of time.

The clock drive is in fact a most important adjunct of the telescope when used for other purposes than ordinary visual observations, though for simple seeing a smooth working slow motion in R. A. answers well. The driving clock from the horological view-point is rudimentary. It consists essentially of a weight-driven, or sometimes spring-driven, drum, turning by a simple gear connection a worm which engages a carefully cut gear wheel on the polar axis, while prevented from running away by gearing up to a fast running fly-ball governor, which applies friction to hold the clockwork down to its rate if the speed rises by a minute amount. There is no pendulum in the ordinary sense, the regularity depending on the uniformity of the total friction—that due to the drive plus that applied by the governor.

Figure 134 shows a simple and entirely typical driving clock by Warner & Swasey. Here A is the main drum with its winding gear at B, C is the bevel gear, which is driven from another carried by A, and serves to turn the worm shaft D; E marks the fly balls driven by the multiplying gearing plainly visible. The governor acts at a predetermined rotation speed to lift the spinning friction disc F against its fixed mate, which can be adjusted by the screw G.

The fly-balls can be slightly shifted in effective position to complete the regulation. These simple clocks, of which there are many species differing mainly in the details of the friction device, are capable of excellent precision if the work of driving the telescope is kept light.

For large and heavy instruments, particularly if used for photographic work where great precision is required, it is difficult to keep the variations of the driving friction within the range of compensation furnished by the governor friction alone, and in such case recourse is often taken to constructions in which the fly balls act as relay to an electrically controlled brake, or in which the driving power is supplied by an electric motor suitably governed either continuously or periodically. For such work independent hand guiding mechanism is provided to supplement the clockwork. For equatorials of the smallest sizes, say 3 to 4 inches aperture, spring operated driving clocks are occasionally used. The general plan of operation is quite similar to the common weight driven forms, and where the weights to be carried are not excessive such clocks do good work and serve a very useful purpose.

An excellent type of the simple spring driving clock is shown in Fig. 136 as constructed by Zeiss. Here 1 is the winding gear,

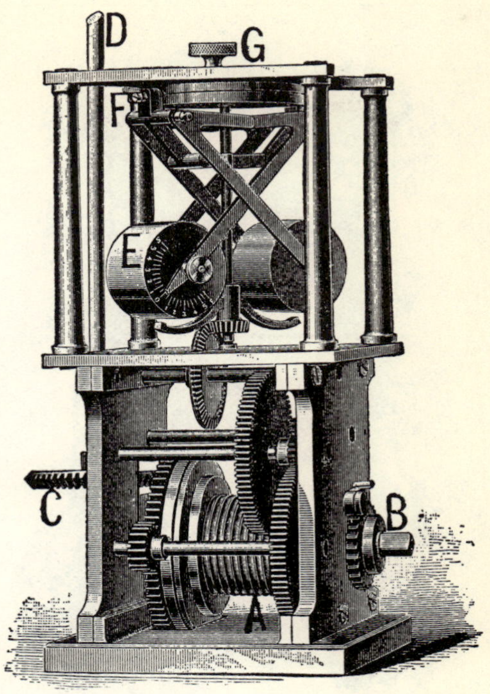

Fig. 134.—Typical Driving Clock. (*Courtesy of The Clarendon Press.*)

2 the friction governor, and 3 the regulating gear. It will be seen that the friction studs are carried by the fly balls themselves, somewhat as in Fraunhofers' construction a century since, and the regulation is very easily and quickly made by adjusting the height of the conical friction surface above the balls.

For heavier work the same makers generally use a powerful weight driven train with four fly-balls and electric seconds control, sometimes with the addition of electric motor slow motions to adjust for R. A. in both directions.

Figure 135 is a rather powerful clock of analogous form by the Clarks. It differs a little in its mechanism and especially in the friction gear in which the bearing disc is picked up by a delicately set latch and carried just long enough to effect the regulation. It is really remarkable that clockworks of so simple character as these should perform as well as experience

FIG. 135.—Clark Driving Clock.

shows that they do. In a few instances clocks have depended on air-fans for their regulating force, something after the manner of the driving gear of a phonograph, but though rather successful for light work they have found little favor in the task of driving equatorials. An excellent type of a second genus is the pendulum controlled driving clock due to Sir David Gill. This has a powerful weight-driven train with the usual fly-ball governor. But the friction gear is controlled by a contact-making seconds pendulum in the manner shown diagrammatically in Fig. 137. Two light leather tipped rods each controlled by an electro magnet act upon an auxiliary brake disc carried by the governor spindle which is set for normal speed with one brake rod bearing lightly on it. Exciting the corresponding magnet relieves the pressure and accelerates the clock, while exciting the other adds braking effect and slows it.

FIG. 136.—Spring Operated Driving Clock.

In Fig. 137 is shown from the original paper, (M. N. Nov., 1873), the very ingenious selective control mechanism. At P is suspended the contact-making seconds-pendulum making momentary contact by the pin Q with a mercury globule at R. Upon a spindle of the clock which turns once a second is fixed a vulcanite disc γ, δ, ϵ, σ. This has a rim of silver broken at the points γ, δ, ϵ, σ, by ivory spacers covering 3° of circumference. On each side of this disc is another, smaller, and with a complete silver rim. One, $\eta\theta$, is shown, connected with the contact spring V; its mate $\eta'\theta'$, on the other side contacts with U, while a third contact K bears on the larger disc.

The pair of segments σ, γ, and δ, ϵ, are connected to η θ, the other pair of segments to $\eta'\theta'$. Now suppose the discs turning with the arrows: If K rests on one of the insulated points when the pendulum throws the battery C Z into circuit nothing happens. If the disc is gaining on the pendulum, K, instead of resting on

γ as shown will contact with segment γ, σ, and actuate a relay via V, exciting the appropriate brake magnet.

If the disc is losing, K contacts with segment γ, δ, and current will pass via $\eta'\theta'$ and U to a relay that operates the other

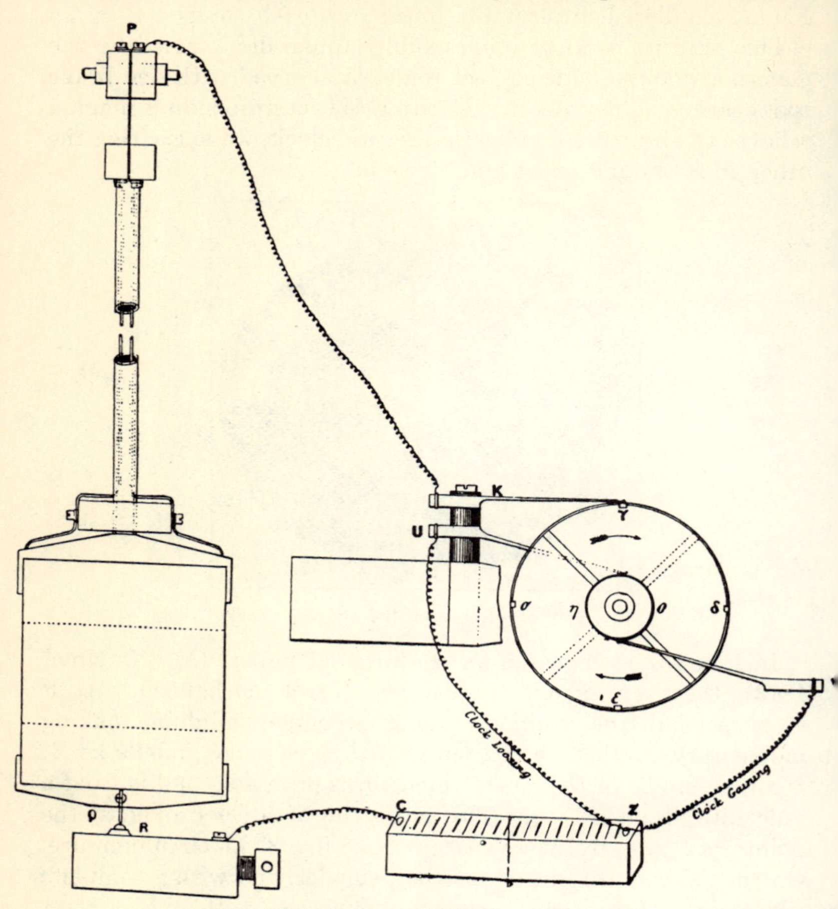

FIG. 137.—Sir David Gill's Electric Control.

brake magnet and lets the clock accelerate. A fourth disc (not shown) on the same spindle is entirely insulated on its edge except at points corresponding to γ and ϵ, and with a contact spring like K.

If the disc is neither gaining nor losing when the pendulum makes contact, current flows via this fourth disc and sets the relay on the mid-point ready to act when needed. This clock is the prototype of divers electrically-braked driving clocks with pendulum control, and proved beautifully precise in action, like various kindred devices constructed since, though the whole genus is somewhat expensive and intricate.

The modern tendency in driving apparatus for telescopes, particularly large instruments, is to utilize an electric motor for the source of power, using a clock mechanism merely for the purpose of accurately regulating the rate of the motor. We thus have the driving clock in its simplest form as a purely mechanical device worked by a sensitive fly-ball governor. The next important type is that in which the clock drive is precisely regulated by a pendulum clock, the necessary governing power being applied electrically as in Fig. 137 or sometimes mechanically.

Finally we come to the type now under consideration where the instrument itself is motor driven and the function of the clock is that of regulating the motor. A very good example of such a drive is the Gerrish apparatus used for practically all the instruments at the various Harvard observatory stations, and which has proved extremely successful even for the most trying work of celestial photography. The schematic arrangement of the apparatus is shown in Fig. 138. Here an electric motor shown in diagram in 1, Fig. 138, is geared down to approximately the proper speed for turning the right ascension axis of the telescope. It is supplied with current either from a battery or in practice from the electric supply which may be at hand. This motor is operated on a 110 volt circuit which supplies current through the switch 2 which is controlled by the low voltage clock circuit running through the magnet 3. The clock circuit can be closed and opened at two points, one controlled by the seconds pendulum 5, the other at 7 by the stud on the timing wheel geared to the motor for one revolution per second. There is also a shunt around the pendulum break, closed by the magnet switch at 6. This switch is mechanically connected to the switch 2 by the rod 4, so that the pair open and close together.

The control operates as follows: Starting with the motor at rest, the clock circuit is switched on, switches 2, 6 being open and 7 closed. At the first beat of the pendulum 2, 6 closes and the current, shunted across the loop containing 5, holds 2 closed until the motor has started and broken the clock circuit at the timer. The fly-wheel carries on until the pendulum again closes the power circuit via 2, 6, and current stays on the motor until the timer has completed its revolution.

This goes on as the motor speeds up, the periodic power supply being shortened as the timer breaks it earlier owing to the acceleration, until the motor comes to its steady speed at which the power is applied just long enough to maintain uniformity. If the motor for any cause tends to overspeed the cut-off is earlier, while slowing down produces a longer power-period bringing the

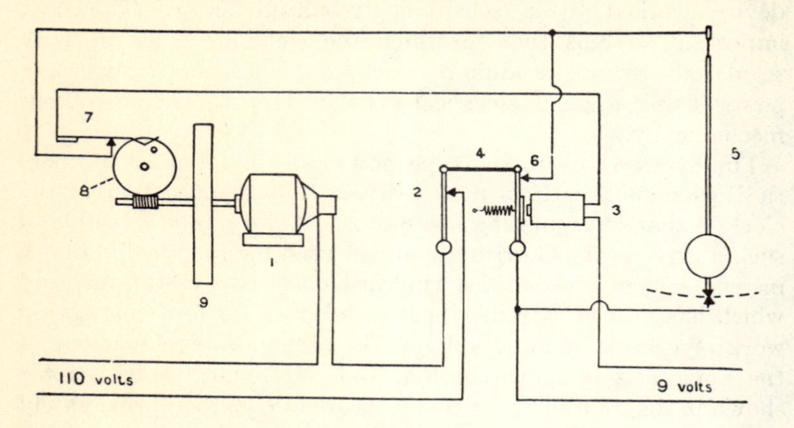

FIG. 138.—Diagram of Gerrish Electric Control.

speed back to normal. The power period is generally ¼ to ½ second. The power supplied to the motor is very small even in the example here shown, only 1 ampere at 110 volts.

The actual proportion of a revolution during which current is supplied the motor is therefore rigorously determined by the clock pendulum, and the motor is selected so that its revolutions are exactly timed to this clock pendulum which has no work to do other than the circuit closing, and can hence be regulated to keep accurate time. The small fly-wheel (9), the weight of which is carefully adjusted with respect to the general amount of work to be done, attached to the motor shaft, effectively steadies its action during the process of government. This Gerrish type has been variously modified in detail to suit the instruments to which it has been applied, always following however the same fundamental principles.

An admirable example of the application of this drive is shown in Fig. 139, the 24 inch reflector at the Harvard Observatory. The mount is a massive open fork, and the motor drive

FIG. 139.—Gerrish Drive on 24-inch Reflector.

is seen on the right of the mount. There are here two motors, ordinary fan motors in size. The right hand motor carries the fly-wheel and runs steadily on under the pendulum control. The other, connected to the same differential gear as the driving motor, serves merely for independent regulation and can be run in either direction by the observer to speed or slow the motion in R. A. These examples of clock drive are merely typical of those which have proved to be successful in use for various service, light and heavy. There are almost innumerable variations on clocks constructed on one or another of the general lines here indicated, so many variations in fact that one almost might say there are few driving clocks which are not in some degree special.

The tendency at present is for large instruments very distinctly toward a motor-driven mechanism operating on the right ascension axis, and governed in one of a considerable variety of ways by an actual clock pendulum. For smaller instruments the old mechanical clock, often fitted with electric brake gear and now and then pendulum regulated, is capable of very excellent work.

The principle of the spectroscope is rudimentarily simple, in the familiar decomposition of white light into rainbow colors by a prism. One gets the phenomena neatly by holding a narrow slit in a large piece of cardboard at arms length and looking at it through a prism held with its edge parallel to the slit. If the light were not white but of a mixture of definite colors each color present would be represented by a separate image of the slit instead of the images being merged into a continuous colored band.

With the sun as source the continuous spectrum is crossed by the dark lines first mapped by Fraunhofer, each representing the absorption by a relatively cool exterior layer of some substance that at a higher temperature below gives a bright line in exactly the same position.

The actual construction of the astronomical spectroscope varies greatly according to its use. In observations on the sun the distant slit is brought nearer for convenience by placing it in the focus of a small objective pointed toward the prisms (the collimator) and the spectrum is viewed by a telescope of moderate magnifying power to disclose more of detail. Also, since there is extremely bright light available, very great dispersion can be used, obtained by several or many prisms, so that the spectrum is both fairly wide, (the length of the slit) and extremely long.

In trying to get the spectrum of a star the source is a point, equivalent to an extremely minute length of a very narrow slit. Therefore no actual slit is necessary and the chief trouble is to get the spectrum wide enough and bright enough to examine.

The simplest form of stellar spectroscope and the one in most common use with small telescopes is the ocular spectroscope arranged much like Fig. 140. This fits into the eye tube of a telescope and the McClean form made by Browning of London consists of an ordinary casing with screw collar *B*, a cylindrical lens *C*, a direct vision prism *c*, *f*, *c*, and an eye-cap *A*.

The draw tube is focussed on the star image as with any other ocular, and the light is delivered through *C* to the prism face nearly parallel, and thence goes to the eye, after dispersion by the prism. This consists of a central prism, *f*, of large angle, made of extremely dense flint, to which are cemented a pair of prisms of light crown *c*, *c*, with their bases turned away from that of *f*.

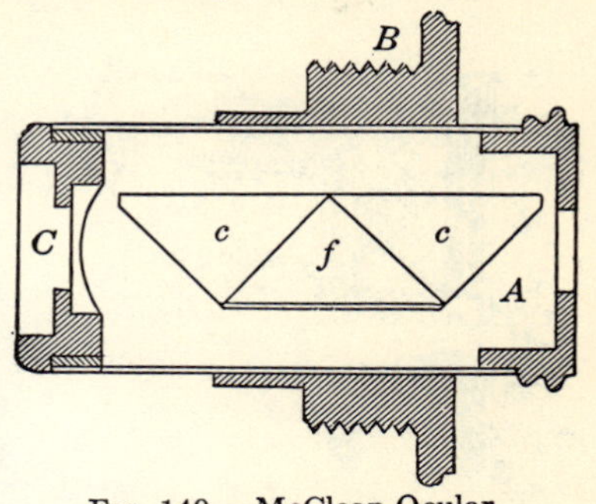

FIG. 140.—McClean Ocular Spectroscope.

We have already seen that the dispersions of glasses vary very much more than their refractions so that with proper choice of materials and angles the refraction of *f* is entirely compensated for some chosen part of the spectrum, while its dispersion quite overpowers that of the crown prisms and gives a fairly long available spectrum.

The cylindrical lens *C* merely serves to stretch out the tiny round star image into a short line thereby giving the resulting spectrum width enough to examine comfortably. The weak cylindrical lens is sometimes slipped over the eye end of the prisms to give the needed width of spectrum instead of putting it ahead of the prisms.

A small instrument of this kind used with a telescope of 3 inches to 5 inches aperture gives a fairly good view of the spectra of stars above second or third magnitude, the qualities of tolerably bright comets and nebulæ and so forth. The visibility of stellar spectra varies greatly according to their type, those with heavy broad bands being easy to observe, while for the same stellar magnitude spectra with many fine lines may be quite beyond examination. Nevertheless a little ocular spectroscope enables one to see many things well worth the trouble of observing.

With the larger instruments, say 6 or 8 inches, one can well take advantage of the greater light to use a spectroscope with a slit, which gives somewhat sharper definition and also an opportunity to measure the spectrum produced.

An excellent type of such an instrument is that shown in Fig. 141, due to Professor Abbé. The construction is analogous to Fig. 140. The ocular is a Huygenian one with the slit mechanism (controlled by a milled head) at A in the usual place of the diaphragm. The slit is therefore in the focus of the eye lens, which serves as collimating lens. Above is the direct vision system J with the usual prisms which are slightly adjustable laterally by the screw P and spring Q.

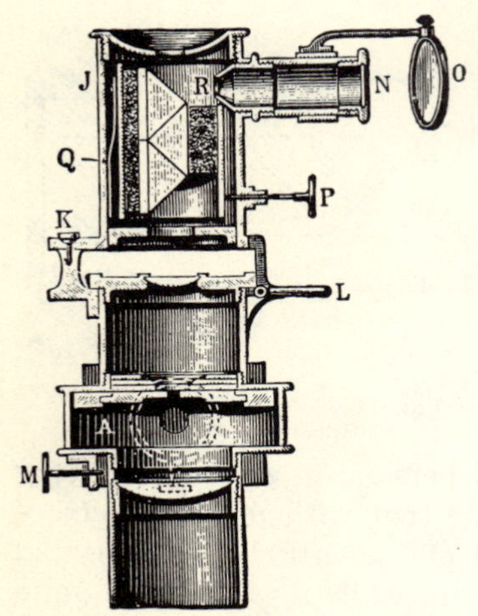

Fig. 141.—Abbé Ocular Spectroscope.

At N is a tiny transparent scale of wave lengths illuminated by a faint light reflected from the mirror O, and in the focus of the little lens R, which transfers it by reflection from the front face of the prism to the eye, alongside the edge of the spectrum. One therefore sees the spectrum marked off by a bright line wave-length scale.

The pivot K and clamp L enable the whole to be swung sidewise so that one can look through the widened slit, locate the star, close the slit accurately upon it and swing on the prisms. M is the clamp in position angle. Sometimes a comparison prism is added, together with suitable means for throwing in spectra of gases or metals alongside that of the star, though these refinements are more generally reserved for instruments of higher dispersion.

To win the advantage of accurate centering of the star in the field gained by the swing-out of the spectroscope in Fig. 141 simple instruments like Fig. 140 are sometimes mounted with an ordinary ocular in a double nose-piece like that used for microscope objectives, so that either can be used at will.

Any ordinary pocket spectroscope, with or without scale or a comparison prism over part of the slit, can in fact be fitted to an adapter and used with the star focussed on the slit and a cylindrical lens, if necessary, as an eye-cap.

Such slit spectroscopes readily give the characteristics of stellar spectra and those of the brighter nebulæ or of comets. They enable one to identify the more typical lines and compare them with terrestrial sources, and save for solar work are about all the amateur observer finds use for.

For serious research a good deal more of an instrument is required, with a large telescope to collect the light, and means for photographing the spectra for permanent record. The cumulative effect of prolonged exposures makes it possible easily to record spectra much too faint to see with the same aperture, and exposures are often extended to many hours.

Spectroscopes for such use commonly employ dense flint prisms of about 60° refracting angle and refractive index of about 1.65, one, two, or three of these being fitted to the instrument as occasion requires. A fine example by Brashear is shown in Fig. 142, arranged for visual work on the 24-inch Lowell refractor. Here A is the slit, B the prism box, C the observing telescope, D the micrometer ocular with electric lamp for illuminating the wires, and E the link motion that keeps the prism faces at equal angles with collimator and observing telescope when the angle between these is changed to observe different parts of the spectrum. This precaution is necessary to maintain the best of definition.

When photographs are to be taken the observing telescope is unscrewed and a photographic lens and camera put in its place. If the brightness of the object permits, three prisms are installed, turning the beam 180° into a camera braced to the same frame alongside the slit.

For purely photographic work, too, the objective prism used by Fraunhofer for the earliest observation of stellar spectra is in wide use. It is a prism fitted in front of the objective with its refracting faces making equal angles with the telescope and the region to be observed, respectively. Its great advantages are small loss of light and the ability to photograph many spectra at once, for all the stars in the clear field of the instrument leave their images spread out into spectra upon the photographic plate.

Figure 143 shows such an objective prism mounted in front of an astrographic objective. The prism is rotatable into any azimuth about the axis of the objective and by the scale i and clamping screw r can have its refracting face adjusted with

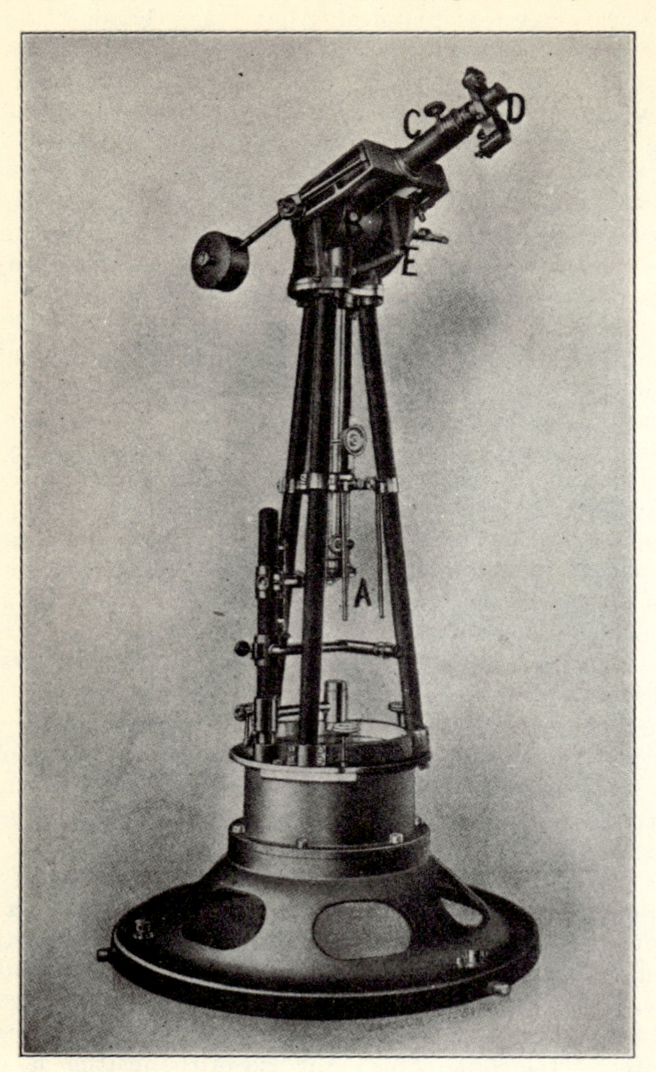

Fig. 142.—Typical Stellar Spectroscope.

respect to that axis to the best position for photographing any part of the spectrum. Such an arrangement is typical of those used for small instruments say from 3 inches to 6 inches aperture.

For larger objectives the prism is usually of decidedly smaller angle, and, if the light warrants high dispersion, several prisms in tandem are used. The objective prism does its best work when applied to true photographic objectives of the portrait lens type which yield a fairly large field. It is by means of big instruments of such sort that the spectra for the magnificent Draper Catalogue have been secured by the Harvard Observatory, mostly at the Arequipa station. In photographing with the objective prism the spectra are commonly given the necessary width for convenient examination by changing just a trifle the rate of the driving clock so that there is a slight and gradual drift in R. A. The refracting edge of the prism being turned parallel to the diurnal motion this drift very gradually and uniformly widens the spectrum to the extent of a few minutes of arc during the whole exposure.

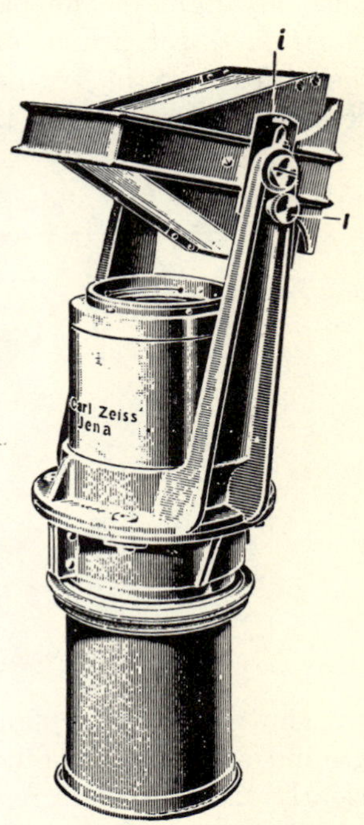

Fig. 143.—Simple Objective Prism.

When one comes to solar spectroscopy one meets an entirely different situation. In stellar work the difficulty is to get enough light, and the tendency is toward large objectives of relatively short focal length and spectroscopes of moderate dispersion. In solar studies there is ample light, and the main thing is to get an image big enough to be scrutinized in detail with very great dispersion.

Especially is this true in the study of the chromospheric flames that rim the solar disc and blaze over its surface. To examine these effectively the spectroscope should have immense dispersion with a minimum amount of stray light in the field to interfere with vision of delicate details.

In using a spectroscope like Fig. 142, if one turned the slit toward the landscape, the instrument being removed from the telescope and the slit opened wide, he could plainly see its various features, refracted through the prism, and appearing in such color as corresponded to the part of the spectrum in the line of the observing telescope. In other words one sees refracted images quite distinctly in spite of the bending of the rays. With high dispersion the image seen is practically monochromatic.

Now if one puts the spectroscope in place, brings the solar image tangent to the slit and then cautiously opens the slit, he sees the bright continuous spectrum of the sky close to the sun, plus any light of the particular color for which the observing telescope is set, which may proceed from the edge of the solar disc. Thus, if the setting is for the red line of hydrogen (C), one sees the hydrogen glow that plays in fiery pillars of cloud about

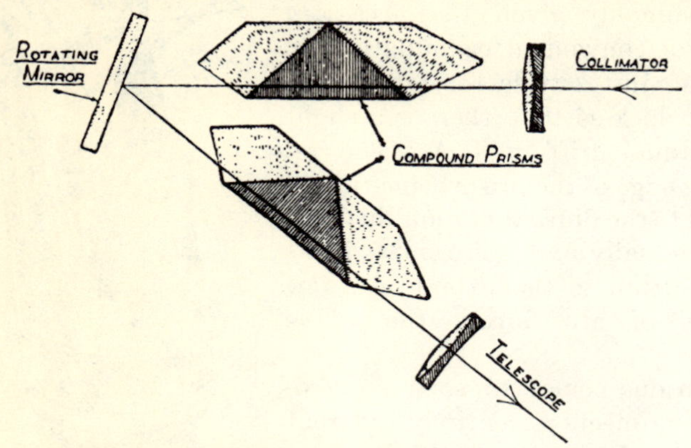

Fig. 144.—Diagram of Evershed Solar Spectroscope.

the sun's limb quite plainly through the opened slit, on a background of light streaming from the adjacent sky. To see most clearly one must use great dispersion to spread this back-ground out into insignificance, must keep other stray light out of the field, and limit his view to the opened slit.

To these ends early solar spectroscopes had many prisms in tandem, up to a dozen or so, kept in proper relation by complicated linkages analogous to the simple one shown in Fig. 142. Details can be found in almost any astronomical work of 40 years ago. They were highly effective in giving dispersion but neither improved the definition nor cut out light reflected back and forth from their many surfaces.

Of late simpler constructions have come into use of which an excellent type is the spectroscope designed by Mr. Evershed and shown in diagram in Fig. 144. Here the path of the rays is from the slit through the collimator objective, then through a very powerful direct vision system, giving a dispersion of 30° through the visible spectrum, then by reflection from the mirror through a second such system, and thence to the observing telescope. The mirror is rotated to get various parts of the spectrum into view, and the micrometer screw that turns it gives means for making accurate measurement of wave lengths.

There are but five reflecting surfaces in the prism system (for the cemented prism surfaces do not count for much) as against more than 20 in one of the older instruments of similar power, and as in other direct vision systems the spectrum lines are substantially straight instead of being strongly curved as with

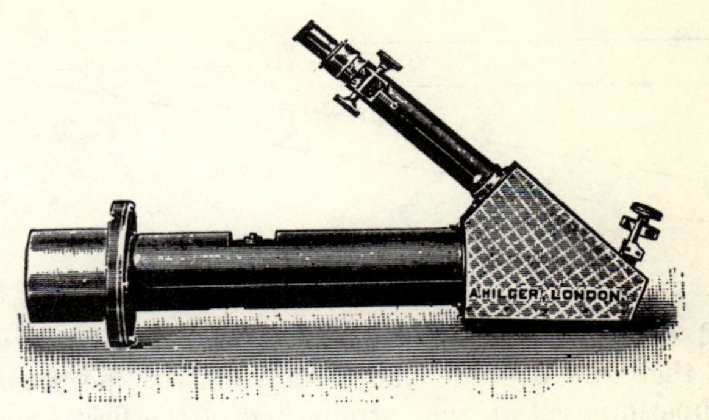

FIG. 145.—Evershed Solar Spectroscope.

multiple single prisms. The result is the light, compact, and powerful spectroscope shown complete in Fig. 145, equally well fitted for observing the sun's prominences and the detailed spectrum from his surface.

In most of the solar spectroscopes made at the present time the prisms are replaced by a diffraction grating. The original gratings made by Fraunhofer were made of wire. Two parallel screws of extremely fine thread formed two opposite sides of a brass frame. A very fine wire was then wound over these screws, made fast by solder on one side of each, and then cut away on the other, so as to leave a grating of parallel wires with clear spaces between.

Today the grating is generally ruled by an automatic ruling engine upon a polished plate of speculum metal. The diamond point carried by the engine cuts very smooth and fine parallel furrows, commonly from 10,000 to 20,000 to the inch. The spaces between the furrows reflect brilliantly and produce diffraction spectra.[1]

When a grating is used instead of prisms the instrument is commonly set up as shown in Fig. 146. Here *A* is the collimator with slit upon which the solar image light falls, *B* is the observing telescope, and *C* the grating set in a rotatable mount with a fine threaded tangent screw to bring any line accurately upon the cross wires of the ocular.

The grating gives a series of spectra on each side of the slit, violet ends toward the slit, and with deviations proportional to

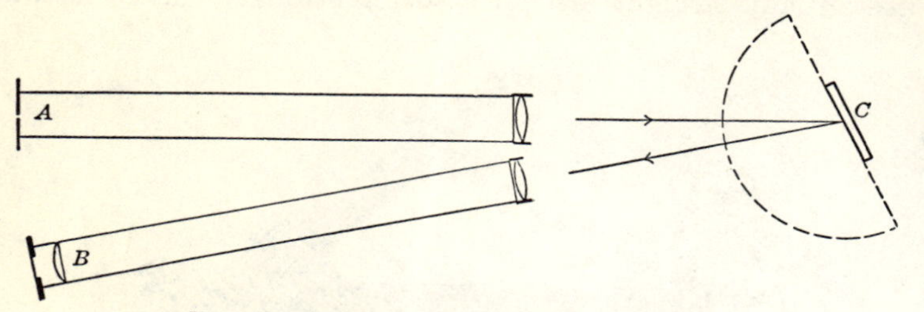

FIG. 146.—Diagram of Grating Spectroscope.

1, 2, 3, 4, etc., times the wave length of the line considered. The spectra therefore overlap, the ultra violet of the second order being superimposed on the extreme red of the first order and so on. Colored screens over the slit or ocular are used to get the overlying spectra out of the way.

The grating spectroscopes are very advantageous in furnishing a wide range of available dispersions, and in giving less stray light than a prism train of equal power. The spectra moreover are very nearly "normal," *i.e.*, the position of each line is proportional to its wave length instead of the blue being disproportionately long as in prismatic spectra.

In examining solar prominences the widened slit of a grating spectroscope shows them foreshortened or stretched to an amount depending on the angular position of the grating, but the effect is easily reckoned.[2]

[1] For the principle of diffraction spectra see Baly, Spectroscopy; Kayser, Handbuch d. Specctroscopie or any of the larger textbooks of physics.

[2] The effect on the observed height of a prominence is $h = h' \frac{\sin c}{\sin t}$, where h is the real height, h′ the apparent height, c the angle made by the grating face with the collimator, and t that with the telescope (Fig. 146).

If the slit is nearly closed one sees merely a thin line, irregularly bright according to the shape of the prominence; a shift of the slit with respect to the solar image shows a new irregular section of the prominence in the same monochromatic light.

These simple phenomena form the basis of one of the most important instruments of solar study—the spectro-heliograph. This was devised almost simultaneously by G. E. Hale and M. Deslandres about 30 years ago, and enables photographs of the sun to be taken in monochromatic light, showing not only the prominences of the limb but glowing masses of gas scattered all over the surface.

The principle of the instrument is very simple. The collimator of a powerful grating spectroscope is provided with a slit the full length of the solar diameter, arranged to slide smoothly on a ball-bearing carriage clear across the solar disc. Just in front of the photographic plate set in the focus of the camera lens is another narrow sliding slit, which, like a focal plane shutter, exposes strip after strip of the plate.

The two slits are geared together by a system of levers or otherwise so that they move at exactly the same uniform rate of speed. Thus when the front slit is letting through a monochromatic section of a prominence on the sun's limb the plate-slit is at an exactly corresponding position. When the front slit is exactly across the sun's center so is the plate slit, at each element of movement exposing a line of the plate to the monochromatic image from the moving front slit. The grating can of course be turned to put any required line into action but it usually is set for the K line (calcium), which is photographically very brilliant and shows bright masses of floating vapor all over the sun's surface.

Figure 147 shows an early and simple type of Professor Hale's instrument. Here A is the collimator with its sliding slit, B the photographic telescope with its corresponding slide and C the lever system which connects the slides in perfectly uniform alignment. The source of power is a very accurately regulated water pressure cylinder mounted parallel with the collimator. The result is a complete photograph of the sun taken in monochromatic light of exactly defined wave length and showing the precise distribution of the glowing vapor of the corresponding substance.

Since the spectroheliograph of Fig. 147, which shows the principle remarkably well, there have been made many modifications, in particular for adapting the scheme to the great horizontal and vertical fixed telescopes now in use. (For details of these see Cont. from the Solar Obs. Mt. Wilson, Nos. 3, 4, 23, and others). The chief difficulty always is to secure entirely smooth and uniform motion of the two moving elements.

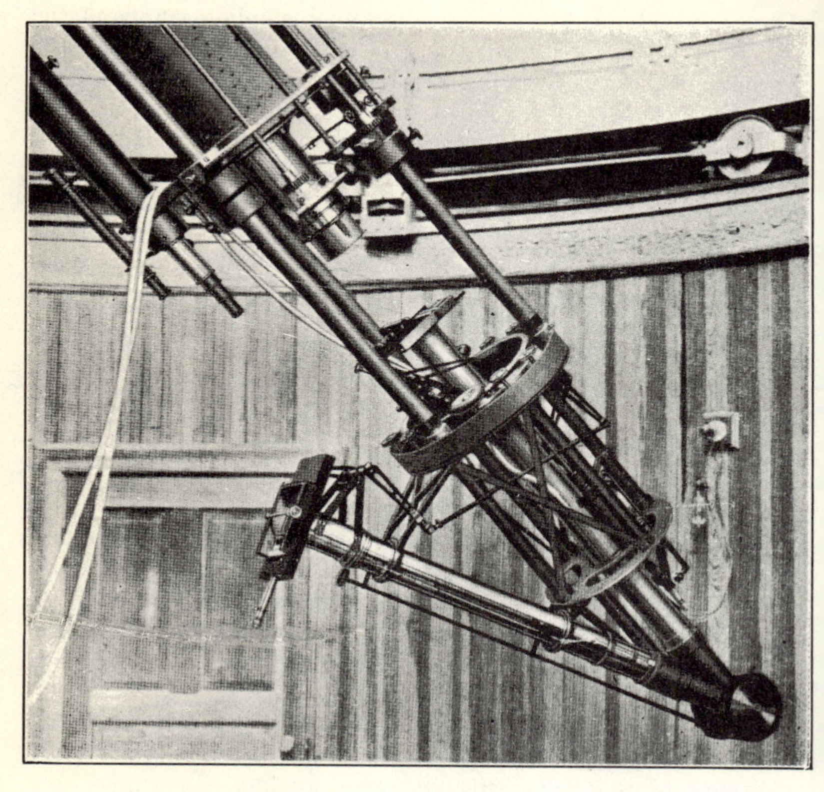

Fig. 147.—Hale's Spectroheliograph (Early Form).

So great and interesting a branch of astronomy is the study of variable stars that some form of photometer should be part of the equipment of every telescope in serious use for celestial observation. An immense amount of useful work has been done by Argelander's systematic method of eye observation, but it is far from being precise enough to disclose many of the most important features of variability.

The conventional way of reckoning by stellar magnitudes is conducive to loose measurements, since each magnitude of difference implies a light ratio of which the log is 0.4, *i.e.*, each magnitude is 2.512 times brighter than the following one. As a result of this way of reckoning the light of a star of mag. 9.9 differs from one of mag. 10.0 not by one per cent but by about nine. Hence to grasp light variations of small order one must be able to measure far below 0.1 of a stellar magnitude.

The ordinary laboratory photometer enables one to compare light sources of anywhere near similar color to a probable error

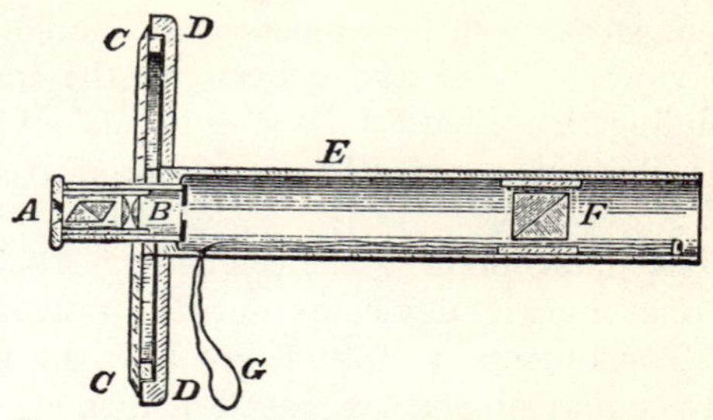

Fig. 148.—Double Image Stellar Photometer.

of well under 0.1 per cent, but it allows a comparison between sharply defined juxtaposed fields from the two illuminants, a condition much more favorable to precision than the comparison of two points of light, even if fairly near together.

Stellar photometers may in principle be divided into three classes. (1) Those in which two actual stars are brought into the same field and compared by varying the light from one or both in a known degree. (2) Those which bring a real star into the field alongside an artificial star, and again bring the two to equality by a known variation, usually comparing two or more stars via the same artificial star; (3) those which measure the light of a star by a definite method of extinguishing it entirely or just to the verge of disappearance in a known progression. Of each class there are divers varieties. The type of the first class may be taken as the late Professor E. C. Pickering's polarizing photometer. Its optical principle is shown in Fig. 148. Here the brightness of two neighboring objects is compared by polarizing at 90° apart the light received from each and reducing the resulting images to equality by an analyzing Nicol prism. The photometer is fully described, with, several other polarizing instruments, in H. A. Vol. 11 from which Fig. 148 is taken.

A is a Nicol prism inserted in the ocular *B*, which revolves carrying with it a divided circle *C* read against the index *D*. In the tube *E* which fits the eye end of the telescope, is placed the double image quartz prism *F* capable of sliding either way without rotation by pulling the cord *G*. With the objects to be compared in the same field, two images of each appear. By turning the analyzing Nicol the fainter image of the brighter can always be reduced to equality with the brighter image of the fainter, and the amount of rotation measures the required ratio of brightness.[1] This instrument works well for objects near enough to be in the same field of view. The distance between the images can be adjusted by sliding the prism *F* back and forth, but the available range of view is limited to a small fraction of a degree in ordinary telescopes.

The meridian photometer was designed to avoid this small scope. The photometric device is substantially the same as in Fig. 148. The objects compared are brought into the field by two exactly similar objectives placed at a small angle so that the images, after passing the double image prism, are substantially in coincidence. In front of each of the objectives is a mirror. The instrument points in the east and west line and the mirrors are at 45° with its axis. One brings Polaris into the field, the other by a motion of rotation about the telescope axis can bring any object in or close to the meridian into the field alongside Polaris. The images are then compared precisely as in the preceding instance.[2] There are suitable adjustments for bringing the images into the positions required.

The various forms of photometer using an artificial star as intermediary in the comparison of real stars differ chiefly in the

[1] If A be the brightness of one object and B that of the other, α the reading of the index when one image disappears and β the reading when the two images are equal then $\frac{A}{B} = \tan^2 (\alpha - \beta)$. There are four positions of the Nicol, 90° apart, for which equality can be established, and usually all are read and the mean taken. (H. A. 11, 1.)

[2] For full description and method see H. A. Vol. 14, also Miss Furness' admirable "Introduction to the Study of Variable Stars," p. 122, et seq. Some modifications are described in H. A. Vol. 23. These direct comparison photometers give results subject to some annoying small corrections, but a vast amount of valuable work has been done with them in the Harvard Photometry.

method of varying the light in a determinate measure. Rather the best known is the Zöllner instrument shown in diagram in Fig. 149. Here *A* is the eye end of the main telescope tube. Across it at an angle of 45° is thrown a piece of plane parallel glass *B* which serves to reflect to the focus the beam from down the side tube, *C*, forming the artificial star.

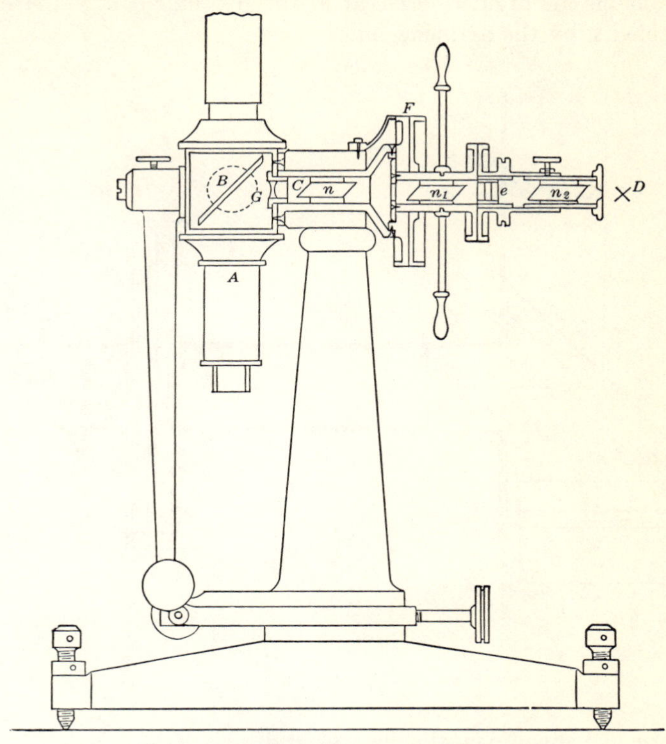

Fig. 149.—Zollner Photometer Diagram.

At the end of this tube is a small hole or more often a diaphragm perforated with several very small holes any of which can be brought into the axis of the tube. Beyond at *D*, is the source of light, originally a lamp flame, now generally a small incandescent lamp, with a ground glass disc or surface uniformly to diffuse the light.

Within the tube *C* lie three Nicol prisms n, n_1, n_2. Of these n, is fixed with respect to the mirror B and forms the analyser, which n_1 and n_2 turn together forming the polarizing system.

Between n_1 and n_2 is a quartz plate e cut perpendicular to the crystal axis. The color of the light transmitted by such a plate in polarized light varies through a wide range. By turning the Nicol n_2 therefore, the color of the beam which forms the artificial star can be made to match the real star under examination, and then by turning the whole system n_2, E, n_1, reading the rotation on the divided circle at F, the real star can be matched in intensity by the artificial one.

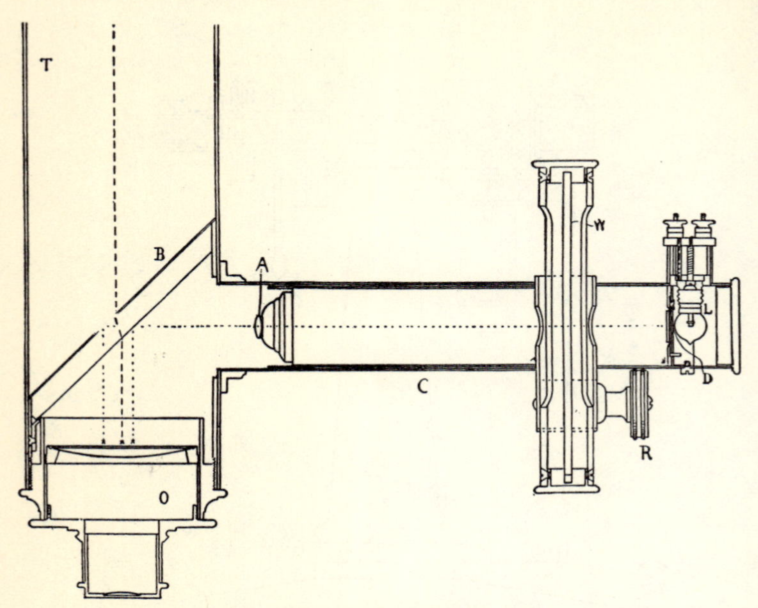

FIG. 150.—Wedge Photometer.

This is viewed via the lens G and two tiny points of light appear in the field of the ocular due respectively to reflection from the front and back of the mirror B, the latter slightly fainter than the former. Alongside or between these the real star image can be brought for a comparison, and by turning the polarizer through an angle α the images can be equalized with the real image. Then a similar comparison is made with a reference star. If A be the brightness of the former and B of the latter then

$$\frac{A}{B} = \frac{\sin^2 \alpha}{\sin^2 \beta}$$

The Zöllner photometer was at first set up as an alt-azimuth instrument with a small objective and rotation in altitude about the axis *C*. Since the general use of electric lamps instead of the inconvenient flame it is often fitted to the eye end of an equatorial.

Another very useful instrument is the modern wedge photometer, closely resembling the Zöllner in some respects but with a very different method of varying the light; devised by the late Professor E. C. Pickering. It is shown somewhat in diagram in Fig. 150. Here as before O is the eye end of the tube, B the plane parallel reflector, C the side tube, L the source of light D the diaphragm and A the lens forming the artificial star by projecting

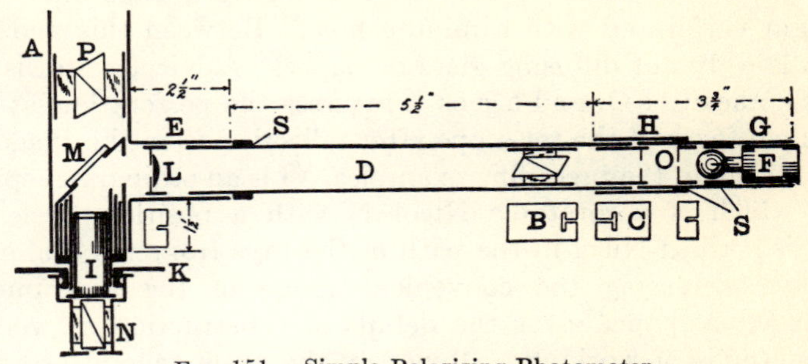

Fig. 151.—Simple Polarizing Photometer.

the hole in the diaphragm. In actual practice the diameter of such hole is 1/100 inch or less.

The light varying device W is a "photographic wedge" set in a frame which is graduated on the edge and moved in front of the aperture by a rack and pinion at R. There are beside colored and shade glasses for use as occasion requires. The "photographic wedge" is merely a strip of fine grained photographic plate given an evenly graduated exposure from end to end, developed, and sealed under a cover glass. Its absorption is permanent, non-selective as to color, and it can be made to shade off from a barely perceptible density to any required opacity. Sometimes a wedge of neutral tinted glass is used in its stead.

Before using such a "wedge photometer" the wedge must be accurately calibrated by observation of real or artificial stars of known difference in brightness. This is a task demanding much care and is well described, together with the whole instrument by Parkhurst (Ap. J. 13, 249). The great difficulty with all instruments of this general type is the formation of an artificial star the image of which shall very closely resemble the image of the real star in appearance and color.

Obviously either the real or artificial star, or both, may be varied in intensity by wedge or Nicols, and a very serviceable modification of the Zöllner instrument, with this in mind was recently described by Shook (Pop. Ast. **27,** 595) and is shown in diagram in Fig. 151. Here A is the tube which fits the ordinary eyepiece sleeve. E is a side tube into which is fitted the extension D with a fitting H at its outer end into which sets the lamp tube G. This carries on a base plug F a small flash light bulb run by a couple of dry cells. At O is placed a little brass diaphragm perforated with a minute hole. Between this and the lamp is a disc of diffusing glass or paper. A Nicol prism is set a little ahead of O, and a lens L focusses the perforation at the principal focus of the telescope after reflection from the diagonal glass M, as in the preceding examples. I is an ordinary eyepiece over which is a rotatable Nicol N with a position circle K. At P is a third Nicol in the path of the rays from the real star, thereby increasing the convenient range of the instrument. The original paper gives the details of construction as well as the methods of working. Obviously the same general arrangement could be used for a wedge photometer using the wedge on either real or artificial star or both.

The third type of visual photometer depends on reducing the light of the star observed until it just disappears. This plan was extensively employed by Professor Pritchard of Oxford some 40 years ago. He used a sliding wedge of dark glass, carefully calibrated, and compared two stars by noting the point on the wedge at which each was extinguished. A photographic wedge may be used in exactly the same way.

Another device to the same end depends on reducing the aperture of the telescope by a "cat's eye," an iris diaphragm, or similar means until the star is no longer visible or just disappearing. The great objection to such methods is the extremely variable sensitivity of the eye under varying stimulus of light.

The most that can be said for the extinction photometer is that in skillful and experienced hands like Pritchard's it has sometimes given much more consistent readings than would be expected. It is now and then very convenient for quick approximation but by no courtesy can it be considered an instrument of precision either in astronomical or other photometry.[1]

The photometer question should not be closed without referring the reader to the methods of physical photometry as developed by Stebbins, Guthnick and others. The first of these depends on the use of the selenium cell in which the electrical resistance falls on exposure of the selenium to light. The device is not one adapted to casual use, and requires very careful nursing to give the best results, but these are of an order of precision beyond anything yet reached with an astronomical visual photometer. Settings come down to variations of something like 2 per cent, and variations in stellar light entirely escaping previous methods become obvious.

The photoelectric cell depends on the lowering of the apparent electric resistance of a layer of rarified inert gas between a platinum grid and an electrode of metallic potassium or other alkali metal when light falls on that electrode. The rate of transmission of electricity is very exactly proportional to the illumination, and can be best measured by a very sensitive electrometer. The results are extraordinarily consistent, and the theoretical "probable error" is very small, though here, as elsewhere, "probable error" is a rather meaningless term apt to lead to a false presumption of exactness. Again the apparatus is somewhat intricate and delicate, but gives a precision of working if anything a little better than that of the selenium cell, quite certainly below 1 per cent.

Neither instrument constitutes an attachment to the ordinary telescope of modest size which can be successfully used for ordinary photometry, and both require reduction of results to the basis of visual effect.[2] But both offer great promise in detecting

[1] The general order of precision attained by astronomical photometers is shown in the discovery, photographically, by Hertzsprung in 1911, that Polaris, used as a standard magnitude for many years, is actually a variable. Its period is very near to four days, its photographic amplitude 0.17 and its visual amplitude about 0.1, *i.e.*, a variation of $\pm$ 5 per cent in the light was submerged in the observational uncertainties, although once known it was traced out in the accumulated data without great difficulty.

[2] Such apparatus is essentially appurtenant to large instruments only, say of not less than 12 inches aperture and preferably much more. The eye is enormously more sensitive as a detector of radiant energy than any device of human contrivance, and thus small telescopes can be well used for visual photometry, the bigger instruments having then merely the advantage of reaching fainter stars.

minute variations of light and have done admirable work. For a good fundamental description of the selenium cell photometer see Stebbins, Ap. J. **32,** 185 and for the photoelectric method see Guthnick A. N. **196,** 357 also A. F. and F. A. Lindemann, M. N. **39,** 343. The volume by Miss Furness already referred to gives some interesting details of both.

CHAPTER IX

THE CARE AND TESTING OF TELESCOPES

A word at the start concerning the choice and purchase of telescopes. The question of refractors vs. reflectors has been already considered. The outcome of the case depends on how much and how often you are likely to use the instrument, and just what you want it for. For casual observations and occasional use—all that most busy buyers of telescopes can expect—the refractor has a decided advantage in convenience. If one has leisure for frequent observations, and particularly if he can give his telescope a permanent mount, and is going in for serious work, he will do well not to dismiss the idea of a reflector without due deliberation.

In any case it is good policy to procure an instrument from one of the best makers. And if you do not buy directly of the actual maker it is best to deal with his accredited agents. In other words avoid telescopes casually picked up in the optical trade unless you chance to have facilities for thorough testing under competent guidance before purchase. No better telescopes are made than can be had from the best American makers. A few British and German makers are quite in the same class. So few high grade French telescopes reach this country as to cause a rather common, but actually unjust,[1] belief that there are none.

If economy must be enforced it is much wiser to try to pick up a used instrument of first class manufacture than to chance a new one at a low price. Now and then a maker of very ordinary repute may turn out a good instrument, but the fact is one to be proved—not assumed. Age and use do not seriously deteriorate a telescope if it has been given proper care. Some of Fraunhofer's are still doing good service after a century, and occasionally an instrument from one of the great makers comes into the market at a real bargain. It may drift back to the maker for resale, or turn up at any optician's shop, and in any case is better worth looking at than an equally cheap new telescope.

[1] E. g., the beautiful astrographic and other objectives turned out by the brothers Henry.

The condition of the tube and stand cuts little figure if they are mechanically in good shape. Most of the older high grade instruments were of brass, beautifully finished and lacquered, and nothing looks worse after hard usage. It is essential that the fitting of the parts should be accurate and that the focussing rack should work with the utmost smoothness. A fault just here, however, can be remedied at small cost. The mount, whatever its character, should be likewise smooth working and without a trace of shakiness, unless one figures on throwing it away.

As to the objective, it demands very careful examination before a real test of its optical qualities. The objective with its cell should be taken out and closely scrutinized in a strong light after the superficial dust has been removed with a camel's hair brush or by wiping very gently with the soft Japanese "lens paper" used by opticians.

One is likely to find plenty to look at; spots, finger marks, obvious scratches, and what is worse a net-work of superficial scratches, or a surface with patches looking like very fine pitting. These last two defects imply the need of repolishing the affected surface, which means also more or less refiguring. Ordinary brownish spots and finger marks can usually be removed with little trouble.

The layman, so to speak, is often warned never to remove the cell from a telescope but he might as well learn the simpler adjustments first as last. In taking off a cell the main thing is to see what one is about and to proceed in an orderly manner. If the whole cell unscrews, as often is the case in small instruments, the only precaution required is to put a pencil mark on the cell and its seat so that it can be screwed back to where it started.

If as is more usual the cell fits on with three pairs of screws, one of each pair will form an abutment against which its mate pulls the cell. A pencil mark locating the position of the head of each of the pulling screws enables one to back them out and replace them without shifting the cell.

The first inspection will generally tell whether the objective is worth further trouble or not. If all surfaces save the front are in good condition it may pay to send the objective to the maker for repolishing. If more than one surface is in bad shape reworking hardly pays unless the lens can be had for a nominal figure. In buying a used instrument from its original source these precautions are needless as the maker can be trusted to stand back of his own and to put it in first class condition.

However, granted that the objective stands well the inspection for superficial defects, it should then be given a real test for figure and color correction, bearing in mind that objectives, even from first class makers, may now and then show slightly faulty corrections, while those from comparatively unknown sources may now and then turn out well. In this matter of necessary testing old and new glasses are quite on all fours save that one may safely trust the maker with a well earned reputation to make right any imperfections. Cleansing other than dusting off and cautiously wiping with damp and then dry lens paper requires removal of the lenses from their cell which demands real care.

With a promising looking objective, old or new, the first test to be applied is the artificial star—artificial rather than natural since the former stays put and can be used by day or by night. For day use the "star" is merely the bright reflection of the sun from a sharply curved surface—the shoulder of a small round bottle, a spherical flask silvered on the inside, a small silvered ball such as is used for Christmas tree decoration, a bicycle ball, or a glass "alley" dear to the heart of the small boy.

The object, whatever it is, should be set up in the sun against a dark background distant say 40 or 50 times the focal length of the objective to be tested. The writer rather likes a silvered ball cemented to a big sheet of black cardboard. At night a pin hole say 1/32 inch or less in diameter through cardboard or better, tinfoil, with a flame, or better a gas filled incandescent lamp behind it, answers well. The latter requires rather careful adjustment that the projected area of the closely coiled little filament may properly fill the pinhole just in front of it.

Now if one sets up the telescope and focusses it approximately with a low power the star can be accurately centered in the field. Then if the eyepiece is removed, the tube racked in a bit, and the eye brought into the focus of the objective, one can inspect the objective for striæ. If these are absent the field will be uniformly bright all over. Not infrequently however one will see a field like Fig. 152 or Fig. 153. The former is the appearance of a 4-inch objective that the author recently got his eye upon. The latter shows typical striæ of the ordinary sort. An objective of glass as bad as shown in Fig. 152 gives no hope of astronomical usefulness, and should be relegated to the porch of a seashore cottage. Figure 153 may represent a condition practically harmless though undesirable.

The next step is a really critical examination of the focal image. Using a moderately high power ocular, magnifying say 50 to the inch of aperture, the star should be brought to the sharpest focus possible and the image closely examined. If the objective is good and in adjustment this image should be a very small spot of light, perfectly round, softening very slightly in its

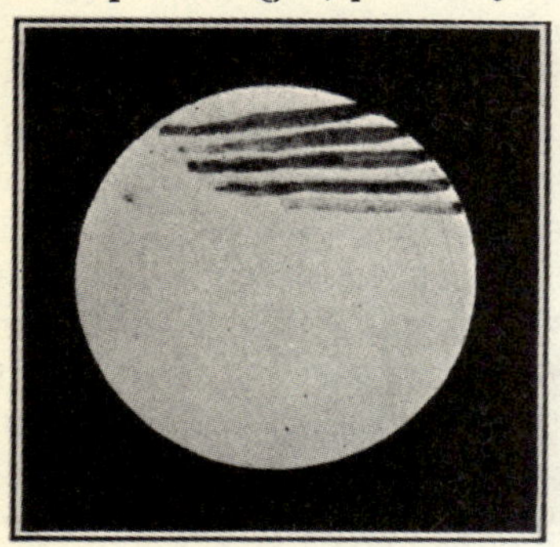

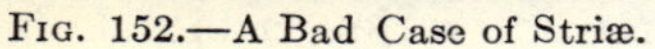

FIG. 152.—A Bad Case of Striæ.

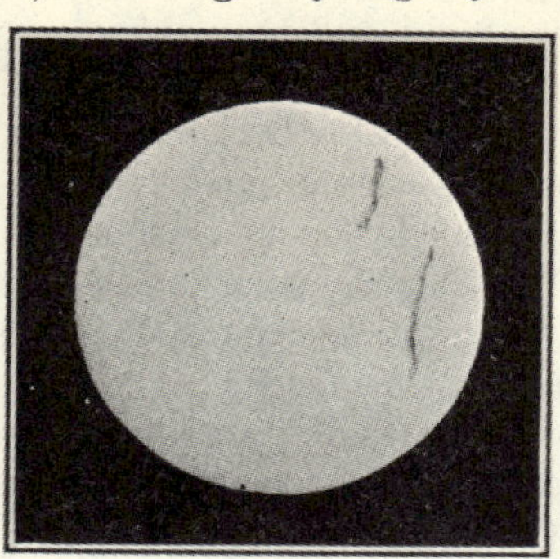

FIG. 153.—Ordinary Striæ.

brilliancy toward the edge, and surrounded by two or three thin, sharp, rings of light, exactly circular and with well defined dark spaces separating them.

Often from the trembling of the air the rings will seem shaky and broken, but still well centered on the star-disc. The general appearance is that shown in Fig. 154.[1]

Instead, several very different appearances may turn up. First, the bright diffraction rings may be visible only on one side of the central disc, which may itself be drawn out in the same direction. Second, the best image obtainable may be fairly sharp but angular or irregular instead of round or oval and perhaps with a hazy flare on one side. Third, it may be impossible to get a really sharp focus anywhere, the image being a mere blob of light with nothing definite about it.

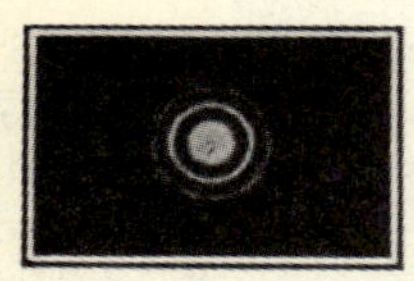

FIG. 154.—A First Class Star Image.

[1] This and several of the subsequent figures are taken from quite the best account of testing objectives: "On the Adjustment and Testing of Telescope Objectives." T. Cooke & Sons, York, 1891, a little brochure unhappily long since out of print. A new edition is just now, 1922, announced.

One should be very sure that the eyepiece is clean and without fault before proceeding further. As to the first point a bit of lens paper made into a tiny swab on a sliver of soft wood will be of service, and the surfaces should be inspected with a pocket lens in a good light to make sure that the cleaning has been thorough. Turning the ocular round will show whether any apparent defects of the image turn with it.

In the first case mentioned the next step is to rack the ocular gently out when the star image will expand into a more or less concentric series of bright interference rings separated by dark spaces, half a dozen or so resulting from a rather small movement out of focus. If these rings are out of round and eccentric like Fig. 155 one has a clear case of failure of the objective to be square with the tube, so that the ocular looks at the image askew.

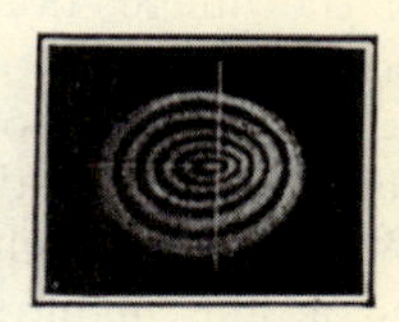

Fig. 155.—Effect of Objective Askew.

In the ordinary forms of objective this means that the side of the objective toward the brighter and less expanded part of the ring system is too near the ocular. This can be remedied by pushing that side of the objective outwards a trifle. Easing off the pulling screw on that side and slightly tightening the abutment screw makes the needed correction, which can be lessened if over done at the first trial, until the ring system is accurately centered. It is a rather fussy job but not at all difficult if one remembers to proceed cautiously and to use the screw driver gently.

In the second case, racking out the ocular a little gives a ring system which exaggerates just the defects of the image. The faults may be due to mechanical strain of the objective in its cell, which is easily cured, or to strains or flaws in the glass itself, which are irremediable. Therefore one should, with the plane of the objective horizontal, loosen the retaining ring that holds the lenses, without disturbing them, and then set it back in gentle contact and try the out-of-focus rings once more. If there is no marked improvement the fault lies in the glass and no more time should be wasted on that particular objective. Fig. 156 is a typical example of this fault.

Fig. 156.—Effect of Flaws in Objective.

In dealing with case three it is well to give the lens a chance by relieving it of any such mechanical strains, for now and then they will apparently utterly ruin the definition, but the prognosis is very bad unless the objective has been most brutally mishandled.

In any case failure to give a sharply defined focus in a very definite plane is a warning that the lens (or mirror) is rather bad. In testing a reflector some pains must be taken at the start with both the main and the secondary mirror. Using an artificial star as before, one should focus and look sharply to the symmetry of the image, taking care to leave the instrument in observing position and screened from the sun for an hour or two before testing. Reflectors are much more sensitive to temperature than refractors and take longer to settle down to stability of figure. With a well mounted telescope of either sort a star at fair altitude on a fine night gives even better testing conditions than an artificial star, (Polaris is good in northern latitudes) but one may have a long wait.

If the reflector is of good figure and well adjusted, the star image, in focus or out, has quite the same appearance as in a refractor except that with a bright star in focus one sees a thin sharp cross of light centered on the image, rather faint but perfectly distinct. This is due to the diffraction effect of the four thin strips that support the small mirror, and fades as the star is put out of focus.

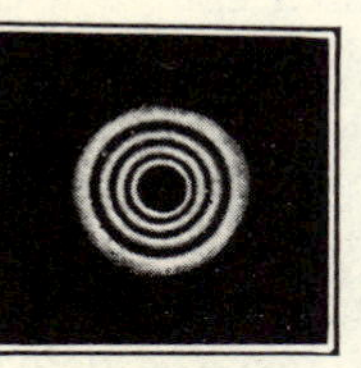

Fig. 157.—Extra-focal Image from Reflector.

The rings then appear as usual, but also a black disc due to the shadowing of the small mirror. Fig. 157 shows the extra-focal image of a real or artificial star when the mirror is well centered, and the star in the middle of the field. There only are the rings round and concentric with the mirror spot. The rings go out of round and the spot out of center for very small departure from the middle of the field when the mirror is of large relative aperture—F/5 or F/6.

If the star image shows flare or oval out-of-focus rings when central of the field, one or both mirrors probably need adjustment. Before laying the trouble to imperfect figure, the mirrors should be adjusted, the small one first as the most likely source of trouble. The side of the mirror toward which the flare or the expanded side of the ring system projects should be slightly pushed away from the ocular. (Note that owing to the reflection this movement is the reverse of that required with a refractor.)

If the lack of symmetry persists one may as well get down to first principles and center the mirrors at once. Perhaps the easiest plan is to prepare a disc of white card-board exactly the size of the mirror with concentric circles laid out upon it and an eighth inch hole in the center. Taking out the ocular and putting a half inch stop in its place one can stand back, lining up the stop with the draw tube, and see whether the small mirror looks perfectly round and is concentric with the reflected circles. If not, a touch of the adjusting screws will be needed.

Then with a fine pointed brush dot the center of the mirror itself through the hole, with white paint. Then, removing the card, one will see this dot accurately centered in the small mirror if the large one is in adjustment, and it remains as a permanent reference point. If the dot be eccentric it can be treated as before, but by the adjusting screws of the large mirror.

The final adjustment can then be made by getting a slightly extra-focal star image fairly in the center of the field with a rather high power and making the system concentric as before described. This sounds a bit complicated but it really is not. If the large mirror is not in place, its counter cell may well be centered and levelled by help of a plumb line from the center of the small mirror and a steel square, as a starting point, the small mirror having been centered as nearly as may be by measurement.[1]

So much for the general adjustment of the objective or mirror. Its actual quality is shown only on careful examination.

As a starting point one may take the extra-focal system of rings given by an objective or mirror after proper centering. If the spherical aberration has thoroughly removed the appearance of the rings when expanded so that six or eight are visible should be like Fig. 158. The center should be a sharply defined bright point and surrounding it, and exactly concentric, should be the

[1] Sometimes with ever so careful centering the ring system in the middle of the field is still eccentric with respect to the small mirror, showing that the axis of the parabola is not perpendicular to the general face of the mirror. This can usually be remedied by the adjusting screws of the main mirror as described, but now and then it is necessary actually to move over the small mirror into the real optical axis. Draper (loc. cit.) gives some experiences of this sort.

interference rings, truly circular and gradually increasing in intensity outwards, the last being very noticeably the strongest.

One can best make the test when looking through a yellow glass screen which removes the somewhat confusing flare due to imperfect achromatism and makes the appearances inside and outside focus closely similar. Just inside or outside of focus the appearance should be that of Fig. 159 for a perfectly corrected objective or mirror.

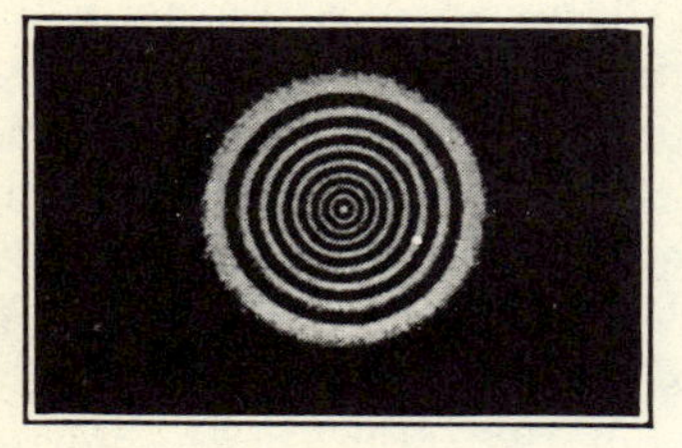

Fig. 158.—Correct Extra-focal Image.

Sometimes an objective will be found in which one edge of the focussed star image is notably red and the opposite one tinted with greenish or bluish, showing unsymmetrical coloring, still more obvious when the image is put a little out of focus. This means that the optical centers of crown and flint are out of line from careless edging of the lenses or very bad fitting. The case is bad enough to justify trying the only remedy available outside the optician's workshop—rotating one lens upon the other and thus trying the pair in different relative azimuths.

The initial positions of the pair must be marked plainly, care must be taken not to displace the spacers 120° apart often found at the edges of the lenses, and the various positions must be tried in an orderly manner. One not infrequently finds a position in which the fault is negligible or disappears altogether, which point should be at once marked for reference.

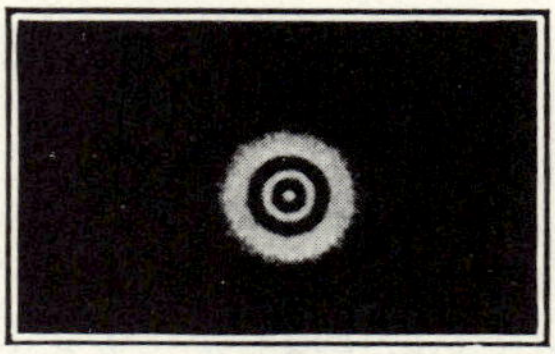

Fig. 159.—Correct Image Just Out of Focus.

In case there is uncorrected spherical aberration there is departure from regular gradation of brightness in the rings. If there is a "short edge," *i.e.*, + spherical aberration, so that rays from the outer zone come to a focus too short, the edge ring will look too strong within focus, and the inner rings relatively weak; with this appearance reversed outside focus. A "long edge" *i.e.*, − spherical aberration, shows the opposite condition, edge rings too strong outside focus and too weak within. Both are rather common faults. The "long edge" effect is shown in Figs. 160 and 161, as taken quite close to focus.

It takes a rather sharp eye and considerable experience to detect small amounts of spherical aberration; perhaps the best way of judging is in quickly passing from just inside to just outside focus and back again, using a yellow screen and watching very closely for variations in brightness. Truth to tell a small amount of residual aberration, like that of Fig. 160, is not a serious matter as regards actual performance—it hurts the telescopist's feelings much more than the quality of his images.

A much graver fault is zonal aberration, where some intermediate zone of objective or mirror comes to a focus too long or too short, generally damaging the definition rather seriously,

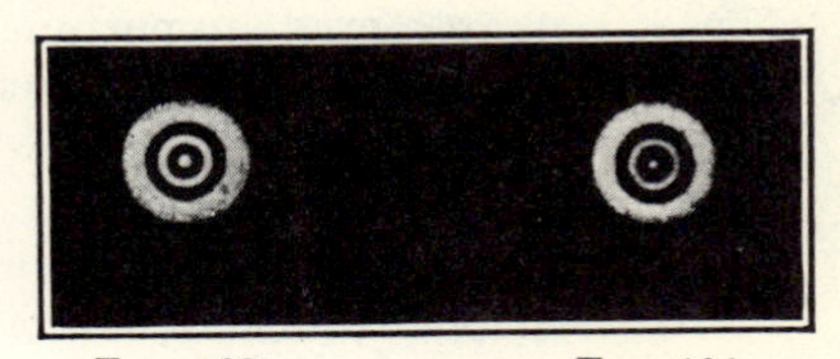

FIG. 160. FIG. 161.
FIG. 160.—Spherical Aberration Just Inside Focus.
FIG. 161.—Spherical Aberration Just Outside Focus.

depending on the amount of variation in focus of the faulty zone. A typical case is shown in Fig. 162 taken within focus. Here two zones are abnormally strong showing, just as in the case of simple spherical aberration, too short focus. Outside of focus the intensities would change places, the outer and midway zones and center being heavy, and the strong zones of Fig. 162 weak. These zonal aberrations are easily detected and are rather common both in objectives and mirrors, though rarely as conspicuous as in Fig. 162.

Another failing is the appearance of astigmatism, which, broadly, is due to a refracting or reflecting surface which is not a surface of revolution and therefore behaves differently for rays incident in different planes around its optical axis. In its commonest form the surface reflects or refracts more strongly along one plane than along another at right angles to it. Hence the two have different foci and there is no point focus at all, but two line foci at right angles. Figs. 163 and 164 illustrate this fault, the former being taken inside and the latter outside focus, under fairly high power. If a star image is oval and the major axis of this oval has turned through 90° when one passes to the other side of focus, astigmatism is somewhere present.

As more than half of humanity is astigmatic, through fault of the eye, one should twist the axis of the eyes some 90° around the axis of the telescope and look again. If the axis of the oval has turned with the eyes a visit to the oculist is in order. If not, it is worth while rotating the ocular. If the oval does not turn with it that particular telescope requires reworking before it can be of much use.

This astigmatism due to fault of figure must not be confused with the astigmatic difference of the image surfaces referred to in Chapter IV which is zero on the axis and not of material importance in ordinary telescopes. Astigmatism of figure on the contrary is bad everywhere and always. It should be especially looked out for in reflecting surfaces, curved or plane, since it is a common result of flexure.

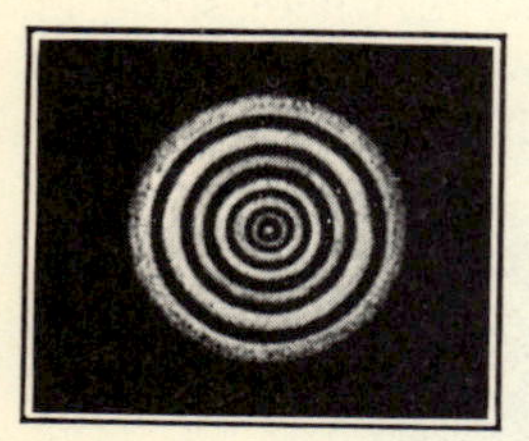

Fig. 162.—A Case of Zonal Aberration.

Passing on now from these simple tests for figure, chromatic aberration has to be examined. Nothing is better than an artificial star formed by the sun in daylight, for the preliminary investigation. At night Polaris is advantageous for this as for other tests.

Fig. 163. Fig. 164.

Fig. 163.—Astigmatism Inside Focus.
Fig. 164.—Astigmatism Outside Focus.

The achromatization curves, Fig. 63, really tell the whole story of what is to be seen. When the telescope is carefully focussed for the bright part of the spectrum, getting the sharpest star image attainable, the central disc, small and clean, should be yellowish white, seen under a power of 60 or 70 per inch of aperture.

But the red and blue rays have a longer focus and hence rim the image with a narrow purplish circle varying slightly in hue according to the character of the achromatization. Pushing the ocular a little inside focus, the red somewhat over-balances the blue and the purple shades toward the red. Pulling out the ocular very slightly one brings the deep red into focus as a minute central red point, just as the image begins to expand a little. Further outside focus a bluish or purplish flare fills the center of the field, while around it lies a greenish circle due to the rays from the middle of the secondary spectrum expanding from their shorter focus.

In an under-corrected objective this red point is brighter and the fringe about the image, focussed or within focus, is conspicuously reddish. Heavy over-correction gives a strong bluish fringe and the red point is dull or absent. With a low power ocular, unless it be given a color correction of its own, any properly corrected objective will seem under-corrected as already explained.

The color correction can also be well examined by using an ocular spectroscope like Fig. 140, with the cylindrical lens removed. Examining the focussed star image thus, the spectrum is a narrow line for the middle color of the secondary spectrum, widening equally at F and B, and expanding into a sort of brush at the violet end. Conversely, when moved outside focus until the width is reduced to a narrow line at F and B, the widening toward the yellow and green shows very clearly the nature and extent of the secondary spectrum. In this way too, the actual foci for the several colors can easily be measured.

The exact nature of the color correction is somewhat a matter of taste and of the uses for which the telescope is designed, but most observers agree in the desirability of the B-F correction commonly used as best balancing the errors of eye and ocular. With reflectors, achromatic or even over-corrected oculars are desirable. The phenomena in testing a telescope for color vary with the class of star observed—the solar type is a good average. Trying a telescope on α Lyrae emphasizes unduly the blue phases, while α Orionis would overdo the red.

The simple tests on star discs in and out of focus here described are ample for all ordinary purposes, and a glass which passes them well is beyond question an admirably figured one. The tests are not however quantitative, and it takes an experienced eye to pick out quickly minor errors, which are somewhat irregular. One sometimes finds the ring system excellent but a sort of haze in the field, making the contrasts poor—bad polish or dirt, but figure good.

A test found very useful by constructors or those with laboratory facilities is the knife edge test, worked out chiefly by Foucault and widely used in examining specula. It consists in principle of setting up the mirror so as to bring the rays to the sharpest possible focus. For instance in a spherical mirror a lamp shining through a pin hole is placed in the centre of curvature, and the reflected image is brought just alongside it where it can be inspected by eye or eyepiece. In Fig. 165 all the rays

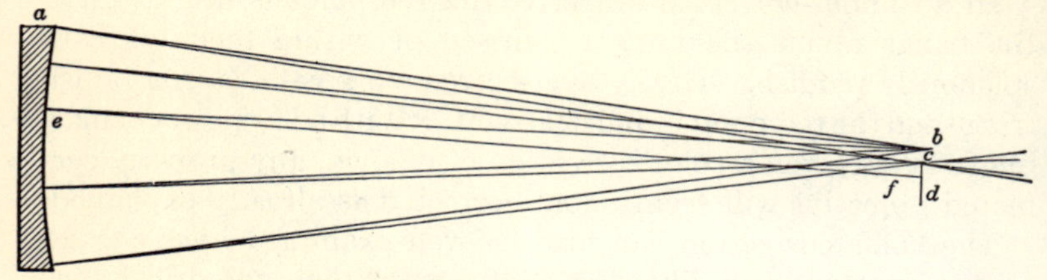

FIG. 165.—The Principle of the Foucault Test.

which emanate from the pinhole *b* and fall on the mirror *a* are brought quite exactly to focus at *c*. The eye placed close to *c* will see, if the mirror surface is perfect, a uniform disc of light from the mirror.

If now a knife edge like *d*, say a safety razor blade, be very gradually pushed through the focus the light will be cut off in

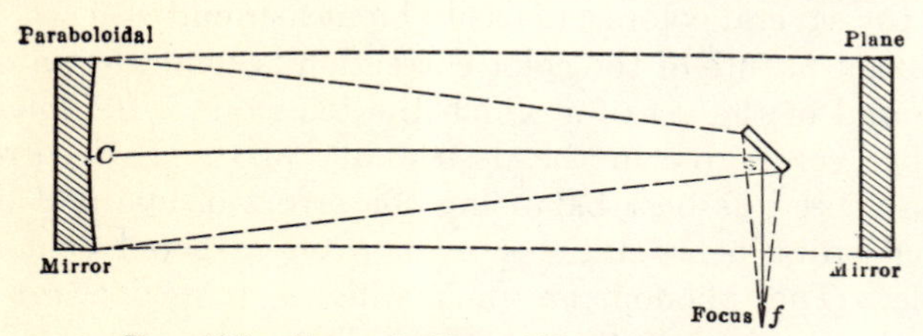

FIG. 166.—Foucault Test of Parabolic Mirror.

a perfectly uniform manner—no zone or local spot going first. If some error in the surface at any point causes the reflected ray to miss the focus and cross ahead of or behind it as in the ray *bef*, then the knife edge will catch it first or last as the case may be, and the spot *e* will be first darkened or remain bright after the light elsewhere is extinguished.

One may thus explore the surface piecemeal and detect not only zones but slight variations in the same zone with great precision. In case of a parabolic mirror as in Fig. 166 the test is made at the focus by aid of the auxiliary plane mirror, and a diagonal as shown, the pinhole and knife edge being arranged quite as before. A very good description of the practical use of the knife edge test may be found in the papers of Dr. Draper and Mr. Ritchey already cited.

It is also applied to refractors, in which case monochromatic light had better be used, and enables the experimenter to detect even the almost infinitesimal markings sometimes left by the polishing tool, to say nothing of slight variations in local figure which are continually lost in the general illumination about the field when one uses the star test in the ordinary manner.

The set-up for the knife edge experiments should be very steady and smooth working to secure precise results, and it therefore is not generally used save in the technique of figuring mirrors, where it is invaluable. With micrometer motions on the knife edge, crosswise and longitudinally, one can make a very exact diagnosis of errors of figure or flexure.

A still more delicate method of examining the perfection of figuring is found in the Hartmann test. (Zeit. fur Inst., 1904, 1909). This is essentially a photographic test, comparing the effect of the individual zones of the objective inside and outside of focus. Not only are the effects of the zones compared but also the effects of different parts of the same zone, so that any lack of symmetry in performance can be at once found and measured.

The Hartmann test is shown diagrammatically in Fig. 167. The objective is set up for observing a natural or artificial star. Just in front of it is placed an opaque screen perforated with holes, as shown in section by Fig. 167, where *A* is the perforated screen. The diameters of the holes are about $\frac{1}{20}$ the diameter of the objective as the test is generally applied, and there are usually four holes 90° apart for each zone. And such holes are not all in one line, but are distributed symmetrically about the screen, care being taken that each zone shall be represented by holes separated radially and also tangentially, corresponding to the pairs of elements in the two astigmatic image surfaces, an arrangement which enables the astigmatism as well as figure to be investigated.

The arrangement of holes actually found useful is shown in Hartmann's original papers, and also in a very important paper by Plaskett (Ap. J. **25** 195) which contains the best account in English of Hartmann's methods and their application. Now

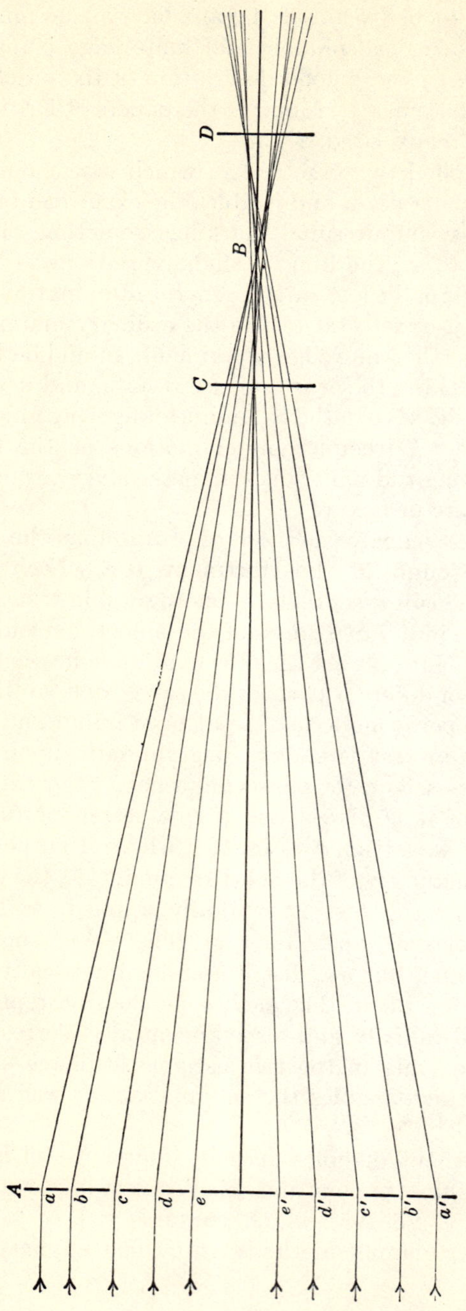

FIG. 167.—The Principle of the Hartmann Test.

each hole in the screen transmits a pencil of light through the objective at the corresponding point, and each pencil comes to a focus and then diverges, the foci being distributed somewhere in the vicinity of what one may regard as the principal focus, *B*. For instance in Fig. 167 are shown five pairs of apertures *a*, *a'*, *b*, *b'*, etc., in five different zones. Now if a photographic plate be exposed a few inches inside focus as at C each pencil from an aperture in the screen will be represented by a dot on the photograph, at such distance from the axis and from the corresponding dot on the other side of the axis as the respective inclinations of the pencils of light may determine.

Similarly a plate exposed at approximately equal distance on the other side of the general focus, as at *D*, will show a pattern of dots due to the distribution of the several rays at a point beyond focus. Now if all the pencils from the several apertures met at a common focus in *B*, the two patterns on the plates *C* and *D* would be exactly alike and for equal distance away from focus of exactly the same size. In general the patterns will not exactly correspond, and the differences measured with the micrometer show just how much any ray in question has departed from meeting at an exact common focus with its fellows.

For instance in the cut it will be observed that the rays *e* and *e'* focus barely beyond *C* and by the time they reach *D* are well spread apart. The relative distance of the dots upon these corresponding plates, with the distance between the plates, shows exactly at what point between *C* and *D* these particular rays actually did cross and come to a focus.

Determining this is merely a matter of measuring up similar triangles, for the path of the rays is straight. Similarly inspection will show that the rays *d* and *d'* meet a little short of *B*, and measurement of their respective records on the plates *C* and *D* would show the existence of a zone intermediate in focus between the focus of *e,e'* and the general focus at *B*. The exact departure of this zone from correct focus can therefore be at once measured.

A little further examination discloses the fact that the outer zone represented by the rays *a,b*, and *a'b'* has not quite the same focus at the two extremities of the same diameter of the objective. In other words the lens is a little bit flatter at one end of this diameter than it is at the other, so that the rays here have considerably longer focus than they should, a fault by no means unknown although fortunately not very common.

It will be seen that the variations between the two screen patterns on *C* and *D*, together with the difference between them, give accurately the performance of each point of the objective represented by an aperture in the screen. And similar investigations by substantially the same method may be extended to the astigmatic variations, to the general color correction, and to the difference in the aberrations for the several colors. The original papers cited should be consulted for the details of applying this very precise and interesting test.

It gives an invaluable record of the detailed corrections of an objective, and while it is one with which the ordinary observer has little concern there are times when nothing else can give with equal precision the necessary record of performance. There are divers other tests used for one purpose or another in examining objectives and mirrors, but those here described are ample for nearly all practical purposes, and indeed the first two commonly disclose all that it is necessary to know.

Now and then one has to deal with an objective which is unmitigatedly dirty. It can be given a casual preliminary cleaning in the way already mentioned, but sometimes even this will not leave it in condition for testing. Then one must get down to the bottom of things and make a thorough job of it.

The chief point to remember in undertaking this is that the thing which one is cleaning is glass, and very easy to scratch if one rubs dust into it, but quite easy to clean if one is careful. The second thing to be remembered is that once cleaned it must be replaced as it was before and not in some other manner.

The possessor of a dirty objective is generally advised to take it to the maker or some reliable optician. If the maker is handy, or an optician of large experience in dealing with telescope objectives is available, the advice is good, but there is no difficulty whatever in cleaning an objective with the exercise of that ordinary care which the user of a telescope may be reasonably expected to possess.

It is a fussy job, but not difficult, and the best advice as to how to clean a telescope objective is to "tub" it, literally, if beyond the stage where the superficial wiping described is sufficient.

To go about the task one first sets down the objective in its cell on a horizontal surface and removes the screws that hold in the retaining ring, or unscrews the ring itself as the case may be. This leaves the cell and objective with the latter uppermost and free to be taken out. Prepare on a table a pad of anything soft, a little smaller than the objective, topping the pad with soft and clean old cloth; then, raising up the cell at an edge, slip the two thumbs under it and lay the fingers lightly on the outer lens of the objective, then invert the whole affair upon the pad and lift off the cell, leaving the objective on its soft bed.

Before anything else is done the edge of the objective should be marked with a hard lead pencil on the edge of both the component lenses, making two well defined v's with their points touching. Also, if, as usual, there are three small separators between the edges of the flint and crown lenses, mark the position of each of these 1, 2, 3, with the same pencil.

Forming another convenient pad of something soft, lift off the upper lens, take out the three separators and lay them in order on a sheet of paper without turning them upside down. Mark alongside each, the serial number denoting its position. Then when these spacers, if in good condition, are put back, they will go back in the same place rightside up, and the objective itself will go back into place unchanged.

Now have at hand a wooden or fibre tub or basin which has been thoroughly washed out with soap and water and wiped dry. Half fill it with water slightly lukewarm and with a good mild toilet soap, shaving soap for example, with clean hands and very soft clean cloth, go at one of the lenses and give it a thorough washing. After this it should be rinsed very thoroughly and wiped dry. As to material for wiping, the main thing is that it must be soft and free from dust that will scratch. Old handkerchiefs serve a good turn.

Dr. Brashear years ago in describing this process recommended cheese cloth. The present day material that goes under this name is far from being as soft at the start as it ought to be, and only the best quality of it should be used, and then only after very thorough soaking, rinsing and drying. The very soft towels used for cleaning cut glass, if washed thoroughly clean and kept free from dust, answer perfectly well. The cheese cloth has the advantage of being comparatively cheap so that it can be thrown away after use. Whatever the cloth, it should be kept, after thorough washing and drying, in a closed jar.

Rinsing the lens thoroughly and wiping it clean and dry is the main second stage of cleansing. It sometimes will be found to be badly spotted in a way which this washing will not remove.

Sometimes the spotting will yield to alcohol carefully rubbed on with soft absorbent cotton or a bunch of lens paper.

If alcohol fails the condition of the surface is such as to justify trying more strenuous means. Nitric acid of moderate strength rubbed on with a swab of absorbent cotton will sometimes clear up the spotting. If this treatment be used it should be followed up with a 10 per cent solution of pure caustic potash or moderately strong c.p. ammonia and then by very thorough rinsing. Glass will stand without risk cautious application of both acid and alkali, but the former better than the latter.

Then a final rinsing and drying is in order. Many operators use a final washing with alcohol of at least 90 per cent strength which is allowed to evaporate with little or no wiping. Alcohol denatured with methyl alcohol serves well if strong enough but beware denatured alcohol of unknown composition. Others have used petroleum naphtha and things of that sort. At the present time these commercial petroleum products are extremely uncertain in quality, like gasoline, being obtained, Heaven knows how, from the breaking down of heavier petroleum products.

If pure petroleum ether can be obtained it answers quite as well as alcohol, but unless the volatile fluid is pure it may leave streaks. Ordinarily neither has to be used, as after the proper wiping the glass comes perfectly clean. This done the glass can be replaced on the pad whence it came and its mate put through the same process.

Flint glass is more liable to spot than the crown, but the crown is by no means immune against the deterioration of the surface, perhaps incipient devitrification, and during the war many objectives "went blind" from unexplained action of this character. As a rule the soap and water treatment applied with care leaves even a pretty hard looking specimen of objective in fairly good condition except for the scratches which previous users have put upon it.

Then if the spacing pieces, usually of tinfoil, are not torn or corroded they can be put back into place, the one lens superimposed upon the other, and the pair put back into the cell by dropping it gently over them and re-inverting the whole, taking care this time to have soft cloth or lens paper under the fingers. Then the retaining ring can be put into place again and the objective is ready for testing or use as the case may be.

If the spacers are corroded or damaged it may be necessary

to replace them with very thin tinfoil cut the same size and shape, leaving however a little extra length to turn down over the edge of the lower lens. They are fastened in place on the extreme edge only by the merest touch of mucilage, shellac or Canada balsam, whichever comes to hand. The one important thing is that the spacers should be entirely free of the sticky material where they lap over the edge of the lens to perform the separation. This lap is generally not over 1/16 of an inch, not enough to show at the outside of the objective when it is in its cell. When the upper lens is lightly pressed down into place, after the gum or shellac is dry, all the projecting portion can be trimmed away with a sharp pen-knife leaving simply the spacers in the appointed places from which the original ones were removed.

Some little space has been given to this matter of cleaning objectives, as in many situations objectives accumulate dirt rather rapidly and it is highly desirable for the user to learn how to perform the simple but careful task of cleansing them.

In ordinary use, when dirt beyond the reach of mere dusting with a camel's hair brush has stuck itself to the exterior of an objective, a succession of tufts of absorbent cotton or wads of lens paper at first dampened with pure water or alcohol and then followed lightly, after the visible dirt has been gently mopped up, by careful wiping with the same materials, will keep the exterior surface in good condition, the process being just that suggested in the beginning of this chapter as the ordinary cleaning up preparatory to a thorough examination.

The main thing to be avoided in the care of a telescope, aside from rough usage generally, is getting the objective wet and then letting it take its chances of drying. In many climates dew is a very serious enemy and the customary dew cap three or four diameters long, bright on the outside and blackened within, is of very great service in lessening the deposit of dew upon the glass. Also the dew cap keeps out much stray light that might otherwise do mischief by brightening the general field. In fact its function as a light-trap is very important especially if it is materially larger in diameter than the objective and provided with stops.

The finder should be similarly protected, otherwise it will mysteriously go blind in the middle of an evening's work due to a heavy deposit of moisture on the objective. The effect is an onset of dimness and bad definition which is altogether obnoxious.

As regards the metal parts of a telescope they should be treated like the metal parts of any other machine, while the moving parts require from time to time a little touch of sperm or similar oil like every other surface where friction may occur.

The old fashioned highly polished and lacquered brass tube was practically impossible to keep looking respectably well provided it was really used to any considerable extent. About the most that could be done to it was dusting when dusty, and cautiously and promptly wiping off any condensed moisture. The more modern lacquered tubes require very little care and if they get in really bad condition can be relacquered without much expense or difficulty.

Wooden tubes, occasionally found in old instruments, demand the treatment which is accorded to other highly finished wooden things, occasional rubbing with oil or furniture polish according to the character of the original surface. Painted tubes may occasionally require a fresh coat, which it does not require great skill to administer. If the surface of wooden tripods comes to be in bad shape it needs the oil or polish which would be accorded to other well finished wooden articles.

Mountings are usually painted or lacquered and either surface can be renewed eventually at no great trouble. Bright parts may be lightly touched with oil as an ordinary rust preventive.

Reflecting telescopes are considerably more troublesome to keep in order than refractors owing to the tender nature of the silvered surface. It may remain in good condition with fairly steady use for several years or it may go bad in a few months or a few weeks. The latter is not an unusual figure in telescopes used about a city where smoke is plentiful. The main thing is to prevent the deposit of dew on the mirror, or getting it wet in any other way, for in drying off the drops almost invariably leave spots.

Many schemes have been proposed for the prevention of injury to the mirror surface. A close fitting metal cover, employed whenever the mirror is not in use, has given good results in many places. Where conditions are extreme this is sometimes lined with a layer of dry absorbent cotton coming fairly down upon the mirror surface, and if this muffler is dry, clean, and a little warmer than the mirror when put on, it seems to be fairly effective. Preferably the mirror should be kept, when not in use, at a little higher temperature than the surrounding air so that dew will not tend to deposit upon it.

As to actual protective measures the only thing that seems to be really efficient is a very thin coating of lacquer, first tried by Perot at the Paris Observatory. The author some ten years since took up the problem in protecting some laboratory mirrors against fumes and moisture and found that the highest grade of white lacquer, such as is used for the coating of fine silverware in the trade, answered admirably if diluted with six or eight volumes of the thinner sold with such commercial lacquers. It is best to thin the lacquer to the requisite amount and then filter.

If now a liberal amount of the mixture is poured upon the mirror surface after careful dusting, swished quickly around, and the mirror is then immediately turned up on edge to drain and dry, a very thin layer of lacquer will be left upon it, only a fraction of a wave length thick, so that it shows broad areas of interference colors.

Treated in this way and kept dry the coating will protect the brilliancy of the silver for a good many months even under rather unfavorable circumstances. After trying out the scheme rather thoroughly the treatment was applied to the 24-inch reflector of the Harvard Observatory and has been in use ever since. The author applied the first coating in the spring of 1913, and since that time it has only been necessary to resilver perhaps once in six months as against about as many weeks previously.

The lacquer used in this case was the so-called "Lastina" lacquer made by the Egyptian Lacquer Company of New York, but there are doubtless others of similar grade in the market. It is a collodion lacquer and in recent years it has proved desirable to use as a thinner straight commercial amylacetate rather than the thinner usually provided with the lacquer, perhaps owing to the fact that difficulty of obtaining materials during the war may have caused, as in so many other cases, substitutions which, while perfectly good for the original purpose did not answer so well under the extreme conditions required in preserving telescope mirrors.

The lacquer coating when thinned to the extent here recommended does not apparently in any way deteriorate the definition as some years of regular work at Harvard have shown. Some experimenters have, however, found difficulty, quite certainly owing to using too thick a lacquer. The endurance of a lacquer coating where the mirror is kept free from moisture, and its power to hold the original brilliancy of the surface is very extraordinary.

The writer took out and tested one laboratory mirror coated seven years before, and kept in a dry place, and found the reflecting power still a little above .70, despite the fact that the coating was so dry as to be almost powdery when touched with a tuft of cotton. At the start the mirror had seen some little use unprotected and its reflection coefficient was probably around .80. If the silver coating is thick as it can be conveniently made, on a well coated mirror, the coat of lacquer, when tarnish has begun, can be washed off with amylacetate and tufts of cotton until the surface is practically clear of it, and the silver itself repolished by the ordinary method and relacquered.

There are many silvering processes in use and which one should be chosen for resilvering a mirror, big or little, is quite largely a matter of individual taste, and more particularly experience. The two most used in this country are those of Dr. Brashear and Mr. Lundin, head of the Alvan Clark Corporation, and both have been thoroughly tried out by these experienced makers of big mirrors.

The two processes differ in several important particulars but both seem to work very successfully. The fundamental thing in using either of them is that the glass surface to be silvered should be chemically clean. The old silver, if a mirror is being resilvered, is removed with strong nitric acid which is very thoroughly rinsed off after every trace of silver has been removed. Sometimes a second treatment with nitric acid may advantageously follow the first with more rinsing. The acid should be followed by a 10 per cent solution of c.p. caustic potash (some operators use c.p. ammonia as easier to clear away) rinsed off with the utmost thoroughness.

On general principles the last rinsing should be with distilled water and the glass surface should not be allowed to dry between this rinsing and starting the silvering process, but the whole mirror should be kept under water until the time for silvering. In Dr. Brashear's process the following two solutions are made up; first the reducing solution as follows:

Rock candy, 20 parts by weight.

Strong nitric acid (spec. gr. 1.22), 1 part.

Alcohol, 20 parts.

Distilled water, 200 parts.

This improves by keeping and if this preparation has to be hurried the acid, sugar and distilled water should be boiled together and then the alcohol added after the solution is cooled.

Second, make up the silvering solution in three distinct portions; first the silver solution proper as follows:

1. 2 parts silver nitrate. 20 parts distilled water.

Second, the alkali solution as follows:

2. 1⅓ parts c.p. caustic potash. 20 parts distilled water.

Third, the reserve silver solution as follows:

3. ¼ part silver nitrate. 16 parts distilled water.

The working solution of silver is then prepared thus: Gradually add to the silver solution No. 1 the strongest ammonia, slowly and with constant stirring. At first the solution will turn dark brown and then it will gradually clear up. Ammonia should be added only just to the point necessary to clear the solution.

Then add No. 2, the alkali solution. Again the mixture will turn dark brown and must be cautiously cleared once more with ammonia until it is straw colored but clear of precipitate. Finally add No. 3, the reserve solution, very cautiously with stirring until the solution grows darker and begins to show traces of suspended matter which will not stir out. Then filter the whole through absorbent cotton to free it of precipitate and it is ready for use. One is then ready for the actual silvering.

Now there are two ways of working the process, with the mirror face up, or face down. The former is advantageous in allowing better inspection of the surface as it forms, and also it permits the mirror of a telescope to be silvered without removing it from the cell, as was in fact done habitually in case of the big reflector of the Alleghany Observatory where the conditions were such as to demand resilvering once a month. The solution was kept in motion during the process by rocking the telescope as a whole.

When silvering face up the mirror is made to form the bottom of the silvering vessel, being fitted with a wrapping of strong paraffined or waxed paper or cloth, wound several times around the rim of the mirror and carried up perhaps half the thickness of the mirror to form a retainer for the silvering solution. This band is firmly tied around the edge of the mirror making a water tight joint. Ritchey uses a copper band fitted to the edge of the mirror and drawn tight by screws, and finishes making tight with paraffin and a warm iron.

In silvering face down the mirror is suspended a little distance above the bottom of a shallow dish, preferably of earthen ware, containing the solution. Various means are used for supporting it. Thus cleats across the back cemented on with hard optician's pitch answer well for small mirrors, and sometimes special provision is made for holding the mirror by the extreme edge in clamps.

Silvering face down is in some respects less convenient but does free the operator from the very serious trouble of the heavy sediment which is deposited from the rather strong silver solution. This is the essential difficulty of the Brashear process in silvering face up. The trouble may be remedied by very gentle swabbing of the surface under the liquid with absorbent cotton, from the time when the silver coating begins fairly to form until it is completed.

The Brashear process is most successfully worked at a temperature between 65° and 70° F. and some experience is required to determine the exact proportion of the reducing solution to be added to the silvering solution. Ritchey advises such quantity of the reducing solution as contains of sugar one-half the total weight of the silver nitrate used. The total amount of solution after mixing should cover the mirror about an inch deep. Too much increases the trouble from sediment and fails to give a clean coating. The requisite quantity of reducing solution is poured into the silvering solution and then immediately, if the mirror is face up, fairly upon it, without draining it of the water under which it has been standing.

If silvering face down the face will have been immersed in a thin layer of distilled water and the mixed solutions are poured into the dish. In either case the solution is rocked and kept moving pretty thoroughly until the process is completed which will take about five minutes. If silvering is continued too long there is likelihood of an inferior whitish outer surface which will not polish well, but short of this point the thicker the coat the better, since a thick coat stands reburnishing where a thin one does not and moreover the thin one may be thin enough to transmit some valuable light.

When the silvering is done the solution should be rapidly poured off, the edging removed or the mirror lifted out of the solution, rinsed off first with tap water and then with distilled, and swabbed gently to clear the remaining sediment. Then the mirror can be set up on edge to dry. A final flowing with alcohol and the use of a fan hastens the process.

In Lundin's method the initial cleaning process is the same but after the nitric acid has been thoroughly rinsed off the surface is gently but thoroughly rubbed with a saturated solution of tin chloride, applied with a wad of absorbent cotton. After the careful rubbing the tin chloride solution must be washed off with the utmost thoroughness, preferably with moderately warm water. It is just as important to get off the tin chloride completely, as it is to clean completely the mirror surface by its use. Otherwise streaks may be left where the silvering will not take well.

When the job has been properly done one can wet the whole surface with a film of water and it will stay wet even when the surface is slightly tilted. As in the Brashear process the mirror must be kept covered with water. Mr. Lundin always silvers large mirrors face up, and forms the dish by wrapping around the edge of the mirror a strip of bandage cloth soaked in melted beeswax and smoothed off by pulling it while still hot between metal rods to secure even distribution of the wax so as to make a water tight joint. This rim of cloth is tied firmly around the edge of the mirror and the strings then wet to draw them still tighter.

Meanwhile the water should cover the mirror by ¾ of an inch or more. It is to be noted that in the Lundin process ordinary water is usually found just as efficient as distilled water, but it is hardly safe to assume that such is the case, without trying it out on a sample of glass.

There are then prepared two solutions, a silver solution,

2.16 parts silver nitrate

100 parts water.

and a reducing solution,

4 parts Merck's formaldehyde

20 parts water.

This latter quantity is used for each 100 parts of the above silver solution, and the whole quantity made up is determined by the amount of liquid necessary to cover the mirror as just described.

The silver solution is cautiously and completely cleared up by strong ammonia as in the Brashear process. The silver and reducing solutions are then mixed, the water covering the mirror poured quickly off, and the silvering solution immediately poured on. The mirror should then be gently rocked and the silver coating carefully watched as it forms.

As the operation is completed somewhat coarse black grains of sediment will form and when these begin to be in evidence the solution should be poured off, the mirror rinsed in running water, the edging removed while the mirror is still rinsing and finally the sediment very gently swabbed off with wet absorbent cotton. Then the mirror can be set up to dry.

The Lundin process uses a considerably weaker silver solution than the Brashear process, is a good deal more cleanly while in action, and is by experienced workers said to perform best at a materially lower temperature than the Brashear process, with the mirror, however, always slightly warmer than the solution. Some workers have had good results by omitting the tin chloride solution and cleaning up the surface by the more ordinary methods. In the Lundin process the solution is sufficiently clear for the density acquired by the silver coating to be roughly judged by holding an incandescent lamp under the mirror. A good coating should show at most only the faintest possible outline of the filament, even of a gas filled lamp.

Whichever process of silvering is employed, and both work well, the final burnishing of the mirror after it is thoroughly dry is performed in the same way, starting by tying up a very soft ball of absorbent cotton in the softest of chamois skin.

This burnisher is used at first without any addition, simply to smooth and condense the film by going over it with quick, short, and gentle circular strokes until the entire surface has been thoroughly cleaned and begins to show a tendency to take polish. Then a very little of the finest optical rouge should be put on to the same, or better another, rubber, and the mirror gone steadily over in a similar way until it comes to a brilliant polish.

A good deal of care should be taken in performing this operation to avoid the settling of dust upon the surface since scratches will inevitably result. Great pains should also be taken not to take any chance of breathing on the mirror or in any other way getting the surface in the slightest degree damp. Otherwise it will not come to a decent polish.

Numerous other directions for silvering will be found in the literature, and all of them have been successfully worked at one time or another. The fundamental basis of the whole process is less in the particular formula used than in the most scrupulous care in cleaning the mirror and keeping it clean until the silvering is completed. Also a good bit of experience is required to enable one to perform the operation so as to obtain a uniform and dense deposit.

At the Harvard Observatory the silvering process in regular use is a modification of the Lundin process devised by E. T. King [Pop. Ast. **30,** 93]. The mirror is cleaned and made ready for silvering as in Lundin's method, except that the stannous chloride is diluted from saturation to half that strength. The mirror is kept covered with water at 65° to 70° F. until ready for the silvering solution. This is double the strength of the Lundin formula and is kept down to a temperature of 45° to 50° F. The exact proportions are

Sol. A Water...100 cc.
Silver nitrate.......................... 4.3 grams.
Add strong ammonia until the precipitate first formed is just redissolved.
Sol. B Water...20 cc.
Formaldehyde (Merck)................ 4 cc.

These, both at 45° to 50° F., are mixed and poured quickly over the mirror from which the wash water has been dashed. The deposit of silver starts at once, the solution turns reddish brown, and in half a minute or so grows muddy with the formation of black grains. When these begin to cling to the mirror surface the process is complete, usually in 3 to 5 minutes. The solution should be quickly poured off, the mirror rinsed, and swabbed with soft cotton in flowing water and set on edge to dry. The burnishing is all important as it raises the reflectivity by fully 20 per cent.

The 24-inch mirrors require a total of about 2 liters of the solutions, and after burnishing are lacquered as already described.

CHAPTER X

SETTING UP AND HOUSING THE TELESCOPE

In regard to getting a telescope into action and giving it suitable protection, two entirely different situations present themselves. The first relates to portable instruments or those on temporary mounts, the second to instruments of position. As respects the two, the former ordinarily implies general use for observational purposes, the latter at least the possibility of measurements of precision, and a mount usually fitted with circles and with a driving clock. Portable telescopes may have either altazimuth or equatorial mounting, while those permanently set up are now quite universally equatorials.

Portable telescopes are commonly small, ranging from about 2½ inches to about 5 inches in aperture. The former is the smallest that can fairly be considered for celestial observations. If thoroughly good and well mounted even this is capable of real usefulness, while the 5-inch telescope if built and equipped in the usual way, is quite the heaviest that can be rated as portable, and deserves a fixed mount.

Setting up an altazimuth is the simplest possible matter. If on a regular tripod it is merely taken out and the tripod roughly levelled so that the axis in azimuth is approximately vertical. Now and then one sets it deliberately askew so that it may be possible to pass quickly between two objects at somewhat different altitudes by swinging on the azimuth axis.

If one is dealing with a table tripod like Fig. 69 it should merely be set on any level and solid support that may be at hand, the main thing being to get it placed so that one may look through it conveniently. This is a grave problem in the case of all small refractors, which present their oculars in every sort of unreachable and uncomfortable position.

Of course a diagonal eyepiece promises a way out of the difficulty, but with small apertures one hesitates to lose the light, and often something of definition, and the observer must pretty nearly stand on his head to use the finder. With well adjusted circles, such are commonly found on a fixed mount, location of objects is easy. On a portable set-up perhaps the easiest remedy is a pair of well aligned coarse sights near the objective end of the tube and therefore within reach when it is pointed zenith-ward. The writer has found a low, armless, cheap splint rocker, such as is sold for piazza use, invaluable under these painful circumstances, and can cordially recommend it.

Even better is an observing box and a flat cushion. The box is merely a coverless affair of any smooth 7/8 inch stuff firmly nailed or screwed together, and of three unequal dimensions, giving three available heights on which to sit or stand. The dimensions originally suggested by Chambers (*Handbook of Astronomy*, II, 215) were 21 × 12 × 15 inches, but the writer finds 18 × 10 × 14 inches a better combination.

The fact is that the ordinary stock telescope tripod is rather too high for sitting, and too low for standing, comfortably. A somewhat stubby tripod is advantageous both in point of steadiness and in accessibility of the eyepiece when one is observing within 30° of the zenith, where the seeing is at its best; and a sitting position gives a much greater range of convenient upward vision than a standing one.

When an equatorial mount is in use one faces the question of adjustment in its broadest aspect. Again two totally different situations arise in using the telescope. First is the ordinary course of visual observation for all general purposes, in which no precise measurements of position or dimensions are involved.

Here exact following is not necessary, a clock drive is convenient rather than at all indispensable, and even circles one may get along without at the cost of a little time. Such is the usual situation with portable equatorials. One does not then need to adjust them to the pole with extreme precision, but merely well enough to insure easy following; otherwise one is hardly better off than with an altazimuth.

In a totally different class falls the instrument with which one undertakes regular micrometric work, or enters upon an extended spectroscopic program or the use of precise photometric apparatus, to say nothing of photography. In such cases a permanent mount is almost imperative, the adjustments must be made with all the exactitude practicable, one finds great need of circles, and the lack of a clock drive is a serious handicap or worse.

Moreover in this latter case one usually has, and needs, some sort of timepiece regulated to sidereal time, without which a right ascension circle is of very little use.

In broad terms, then, one has to deal, first; with a telescope on a portable mount, with or without position circles, generally lacking both sidereal clock and driving clock, and located where convenience dictates; second, with a telescope on a fixed mount in a permanent location, commonly with circles and clock, and with some sort of permanent housing.

Let us suppose then that one is equipped with a 5-inch instrument like Fig. 168, having either the tripod mount, or the fixed pillar mount shown alongside it; how shall it be set up, and, if on the fixed mount, how sheltered?

In getting an equatorial into action the fundamental thing is to place the optical axis of the telescope exactly parallel to the polar axis of the mount and to point the latter as nearly as possible at the celestial pole.

The conventional adjustments of an equatorial telescope are as follows:

1. Adjust polar axis to altitude of pole.
2. Adjust index of declination circle.
3. Adjust polar axis to the meridian.
4. Adjust optical axis perpendicular to declination axis.
5. Adjust declination axis perpendicular to polar axis.
6. Adjust index of right ascension circle, and
7. Adjust optical axis of finder parallel to that of telescope.

Now let us take the simplest and commonest case, the adjustment of a portable equatorial on a tripod mount, when the instrument has a finder but neither circles nor driving clock. Adjustments 2 and 6 automatically drop out of sight, 5 vanishes for lack of any means to make the adjustment , and on a mount made with high precision, like the one before us, 4 is negligible for any purpose to which our instrument is applicable.

Adjustments 1, 3 and 7 are left and these should be performed in the order 7, 1, 3, for sake of simplicity. To begin with the finder has cross-wires in the focus of its eyepiece, and the next step is to provide the telescope itself with similar cross-wires.

These can readily be made, if not provided, by cutting out a disc of cardboard to fit snugly either the spring collar just in front of a positive eyepiece or the eyepiece itself at the diaphragm, if an ordinary Huygenian. Rule two diametral lines on the circle struck for cutting the cardboard, crossing at the center, cut out the central aperture, and then very carefully stretch

FIG. 168.—Clark 5-inch with Tripod and Pier.

over it, guided by the diametral lines, two very fine threads or wires made fast with wax or shellac.

Now pointing the telescope at the most distant well defined object in view, rotate the spring collar or ocular, when, if the crossing of the threads is central, their intersection should stay on the object. If not shift a thread cautiously until the error is corrected.

Keeping the intersection set on the object by clamping the tube, one turns attention to the finder. Either the whole tube is adjustable in its supports or the cross-wires are capable of adjustment by screws just in front of the eyepiece. In either case finder tube or cross-wires should be shifted until the latter bear squarely upon the object which is in line with the cross threads of the main telescope. Then the adjusting screws should be tightened, and the finder is in correct alignment.

As to adjustments 1 and 3, in default of circles the ordinary astronomical methods are not available, but a pretty close approximation can be made by levelling. A good machinist's level is quite sensitive and reliable. The writer has one picked out of stock at a hardware shop that is plainly sensitive to 2′ of arc, although the whole affair is but four inches long.

Most mounts like the one of Fig. 168 have a mark ruled on the support of the polar axis and a latitude scale on one of the cheek pieces. Adjustment of the polar axis to the correct altitude is then made by placing the level on the declination axis, or any other convenient place, bringing it to a level, and then adjusting the tripod until the equatorial head can be revolved without disturbing this level. Then set the polar axis to the correct latitude and adjustment number 1 is complete for the purpose in hand.

Lacking a latitude scale, it is good judgment to mark out the latitude by the help of the level and a paper protractor. To do this level the polar axis to the horizontal, level the telescope tube also, and clamp it in declination to maintain it parallel. Then fix the protractor to a bit of wood tied or screwed to the telescope support, drop a thin thread plumb line from a pin driven into the wood, the declination axis being still clamped. note the protractor reading, and then raise the polar axis by the amount of the latitude.

Next, with a knife blade scratch a conspicuous reference line on the sleeve of the polar axis and its support so that when the equatorial head is again levelled carefully you can set approximately to the latitude at once.

Now comes adjustment 3, the alignment of the polar axis to the meridian. One can get it approximately by setting the telescope tube roughly parallel with the polar axis and, sighting along it, shifting the equatorial head in azimuth until the tube points to the pole star. Then several methods of bettering the adjustment are available.

At the present date Polaris is quite nearly 1° 07′ from the true pole and describes a circle of that radius about it every 24 hours. To get the correct place of the pole with reference to Polaris one must have at least an approximate knowledge of the place of that star in its little orbit, technically its hour-angle. With a little knowledge of the stars this can be told off in the skies almost as easily as one reckons time on a clock. Fig. 169 is, in

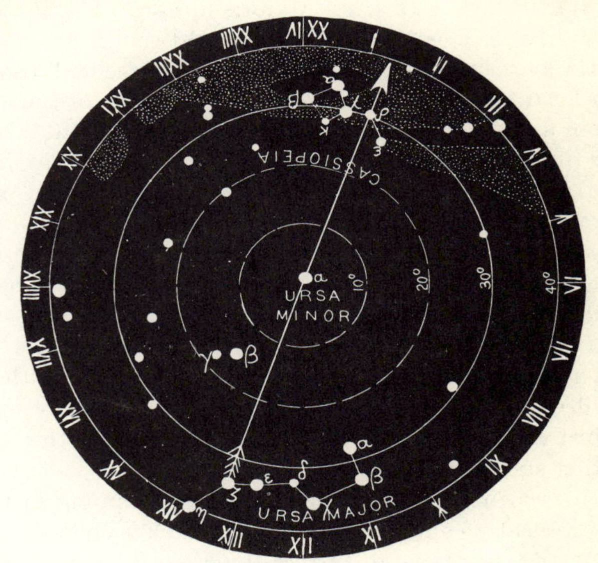

FIG. 169.—The Cosmic Clock.

fact, the face of the cosmic clock, with a huge sweeping hour hand that he who runs may read.

It is a clock in some respects curious; it has a twenty-four hour face like some clocks and watches designed for Continental railway time; the hour hand revolves backward, ("counter-clockwise") and it stands in the vertical not at noon, but at 1.20 Star Time. The two stars which mark the tip and the reverse end of the hour hand are Delta Cassiopeæ and Zeta Ursæ Majoris respectively. The first is the star that marks the bend in the back of the great "chair," the second (Mizar), the star which is next to the end of the "dipper" handle.

One or the other is above the horizon anywhere in the northern hemisphere. Further, the line joining these two stars passes almost exactly through the celestial pole, and also very nearly through Polaris, which lies between the pole and δ Cassiopeæ. Consequently if you want to know the hour-angle of Polaris just glance at the clock and note where on the face δ Cassiopeæ stands, between the vertical which is XXIV o'clock, and the horizontal, which is VI (east) or XVIII (west) o'clock.

You can readily estimate its position to the nearest half hour, and knowing that the great hour hand is vertical (δ Cassiopeæ up) at I^h 20^m or (ζ Ursæ Majoris up) at XIII^h 20^m, you can make a fairly close estimate of the sidereal time.

A little experience enables one to make excellent use of the clock in locating celestial objects, and knowledge of the approximate hour angle of Polaris thus observed can be turned to immediate use in making adjustment 3. To this end slip into the plane of the finder cross-wires a circular stop of metal or paper having a radius of approximately 1° 15′ which means a diameter of 0.52 inch per foot of focal length.

Then, leaving the telescope clamped in declination as it was after adjustment 1, turn it in azimuth across the pole until the pole star enters the field which, if the finder inverts it will do on the other side of the center; i.e. if it stands at IV to the naked eye it will enter the field apparently from the XVI o'clock quarter. When just comfortably inside the field, the axis of the telescope is pointing substantially at the pole.

It is better to get Polaris in view before slipping in the stop and if it is clearly coming in too high or too low shift the altitude of the polar axis a trifle to correct the error. This approximate setting can be made even with the smallest finder and on any night worth an attempt at observation.

With a finder of an inch or more aperture a very quick and quite accurate setting to the meridian can be made by the use of Fig. 170, which is a chart of all stars of 8 mag. or brighter within 1° 30′ of the pole. There are only three stars besides Polaris at all conspicuous in this region, one quite close to Polaris, the other two forming with it the triangle marked on the chart. These two are, to the left, a star of magnitude 6.4 designated B. D. 88° 112, and to the right one of magnitude 7.0, B. D. 89° 13.

The position of the pole for the rest of the century is marked on the vertical arrow and with the stars in the field of the finder one can set the cross-wires on the pole, the instrument remaining clamped in declination, within a very few minutes of arc, quite closely enough for any ordinary use of a portable mount. All this could be done even better with the telescope itself, but it is very rare to find an eyepiece with sufficient field.

At all events the effect of any error likely to be made in these adjustments is not serious for the purpose in hand, since if one

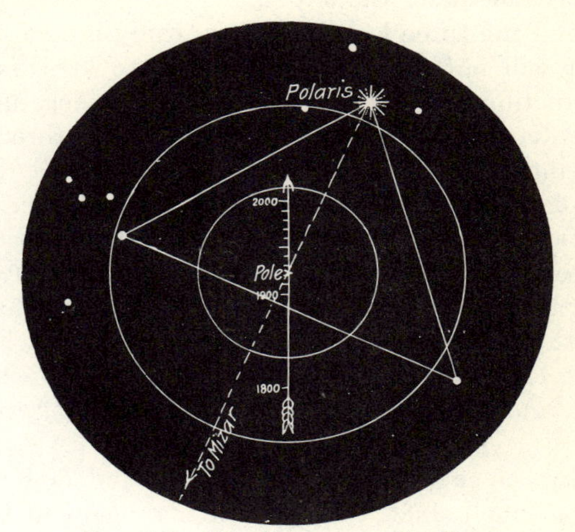

Fig. 170.—The Pole among the Stars.

makes an error of a minute of arc in the setting the resulting displacement of a star in the field will even in the most unfavorable case reach this full amount only after 6 hours following; i.e. with any given eyepiece an error of adjustment equal to the radius of the field will still permit following a star for an hour or two before it drifts inconveniently wide of the center.

Considerable space has been devoted to these easy approximations in setting up, since the directions commonly given require circles and often a clock drive.

In some cases one has to set up a portable equatorial where from necessity for clear sky space, Polaris is not visible. The best plan then is to set up with great care where Polaris can be seen, paying especial attention to the levelling. Then establish two meridian marks on stakes at a convenient distance by turning the telescope 180° on its declination axis and sighting through it in both directions. Now with a surveyor's tape transfer the meridian line East or West as the case may be until it can be used where there is clear sky room.

Few observers near a city can get good sky room, from the interference of houses, trees or blazing street lamps, and the telescope must often be moved from one site to another to reach different fields. In such case it is wise to take the very first step toward giving the telescope a local habitation by establishing a definite placement for the tripod.

To this end the three legs should be firmly linked together by chains that will not stretch—leg directly to leg, and not to a common junction. Then see to it that each leg has a strong and moderately sharp metal point, and, the three points of support being thus definitely fixed, establish the old reliable point-slot-plane bearing as follows:

Lay out at the site (or sites) giving the desired clear view, a circle scratched on the ground of such size that the three legs of your tripod may rest approximately on its periphery. Then lay out on the circle three points 120° apart. At each point sink a short post 12 to 18 inches long and of any convenient diameter, well tarred, and firmly set with the top levelled off quite closely horizontal.

To the top of each bolt a square or round of brass or iron about half an inch thick. The whole arrangement is indicated in diagram in Fig. 171. In *a* sink a conical depression such as is made by drilling nearly through with a 1 inch twist drill. The angle here should be a little broader than the point on the tripod leg. In *b* have planed a V shaped groove of equally broad angle set with its axis pointing to the conical hole in *a*. Leave the surface of *c* a horizontal plane.

Now if you set a tripod leg in *a*, another in the slot at *b* and the third on *c*, the tripod will come in every instance to the same level and orientation. So, if you set up your equatorial carefully in the first place and leave the head clamped in azimuth, you can take it in and replace it at any time still in adjustment as exact as at the start. And if it is necessary to shift from one location to another you can do it without delay still holding accurate adjustment of the polar axis to the pole, and avoiding the need of readjustment.

In case the instrument has a declination circle the original set-up becomes even simpler. One has only to level the tripod, either with or without the equatorial head in place, and then to set the polar axis either vertical or horizontal, levelling the tube with it either by placing the level across the objective cell perpendicular to the declination axis, or laying it along the tube when horizontal.

Then, reading the declination circle, one can set off the co-latitude or latitude as the case may be and, leaving the telescope

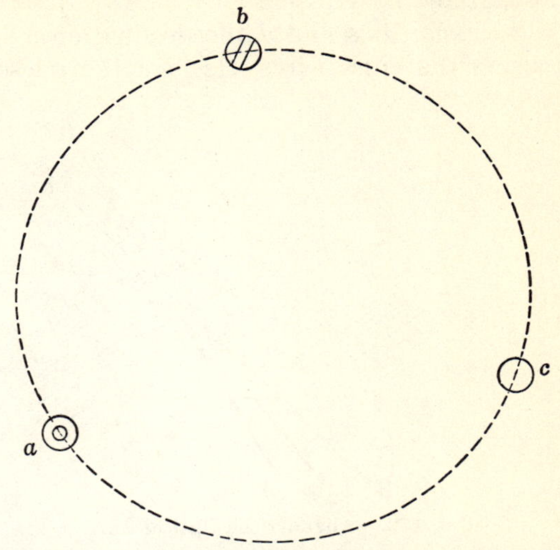

FIG. 171.—A Permanent Foothold for the Tripod.

clamped in declination, lower or raise the polar axis until the tube levels to the horizontal. When the mount does not permit wide adjustment and has no latitude scale one is driven to laying out a latitude templet and, placing a straight edge under the equatorial head, or suspending a plumb line from the axis itself, setting it mechanically to latitude.

Now suppose we are dealing with the same instrument, but are planning to plant it permanently in position on its pillar mount. It is now worth while to make the adjustments quite exactly, and to spend some time about it. The pillar is commonly assembled by well set bolts on a brick or concrete pier. The preliminary steps are as already described.

The pillar is levelled across the top, the equatorial head, which turns upon it in azimuth, is levelled as before, the adjustment being made by metal wedges under the pillar or by levelling screws in the mount if there are any. Then the latitude is set off by the scale, or by the declination circle, and the polar axis turned to the approximate meridian as already described.

There is likely to be an outstanding error of a few minutes of arc which should in a permanent mount be reduced as far as practicable. At the start adjust the declination of the optical axis of the telescope to that of the polar axis. This is done in the manner suggested by Fig. 172.

Here p is the polar axis and d the declination axis. Now if one sights, using the cross-wires, through the telescope a star

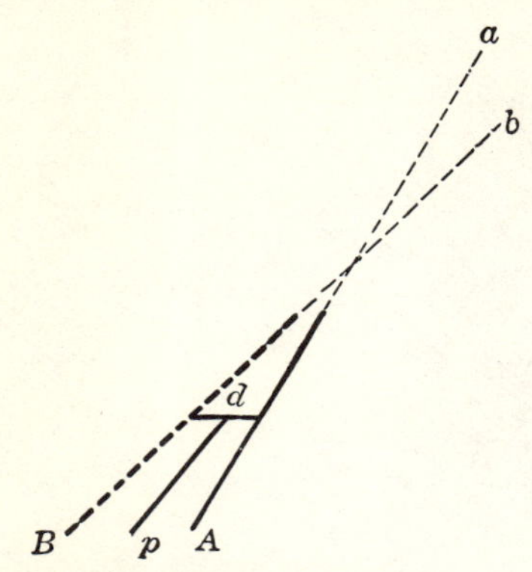

FIG. 172.—Aligning the Optical Axis.

near the meridian, i.e., one that is changing in altitude quite slowly, starting from the position A with the telescope east of the polar axis, and turns it over 180° into the position B, west of the polar axis, the prolongation of the line of sight, b, will fall below a, when as here the telescope points too high in the A position.

In other words the apparent altitude of the star will change by twice the angle between A and p. Read both altitudes on the declination circle and split the difference with the slow motion as precisely as the graduation of the declination circle permits.

The telescope will probably not now point exactly at the star, but as the tube is swung from the A to the B position and back the visible stars will describe arcs of circles which should be nearly concentric with the field as defined by the stop in the eyepiece. If not, a very slight touch on the declination slow motion one way or the other will make them do so to a sufficient exactness, especially if a rather high power eyepiece is used.

The optical axis of the telescope is now parallel to the polar axis, but the latter may be slightly out of position in spite of the preliminary adjustment. Now reverting to the polar field of Fig. 170, swing from position *A* to *B* and back again, correcting any remaining eccentricity of the star arcs around the pole by cautious shifting of the polar axis, leaving the telescope clamped in declination. The first centering is around the pole of the instrument, the second around the celestial pole by help of a half dozen small stars within a half degree on both sides of it, magnitudes 9 and 10, easily visible in a 3-inch or 4-inch telescope, using the larger field of the finder for the coarse adjustment.

If the divided circles read to single minutes or closer, which they generally do not on instruments of moderate size, one can use the readings to set the polar axis and the declination circle, and to make the other adjustments as well.

In default of this help, the declination circle adjustment may be set to read 90° when the optical axis is brought parallel to the polar axis, and after the adjustment of the latter is complete, the R. A. circle can be set by swinging up the telescope in the meridian and watching for the transit of any star of known R. A. over the central cross-wire, at which moment the circle should be clamped to the R. A. thus defined.

Two possible adjustments are left, the perpendicularity of the polar and declination axes, and that of the optical axis to the declination axis. As a rule there is no provision for either of these, which are supposed to have been carried out by the maker. The latter adjustment if of any moment will disclose itself as a lateral wobble in trying to complete the adjustment of optical axis to polar axis. It can be remedied by a liner of tinfoil or even paper under one end of the tube's bearing on its cradle. Adjustment of the former is strictly a job for the maker.

For details of the rigorous adjustments on the larger instruments the reader will do well to consult Loomis' *Practical Astronomy* page 28 and following.[1] The adjustments here considered are those which can be effectively made without driving clock, finely divided circles, or exact knowledge of sidereal time. The first and last of these auxiliaries, however, properly belong with an instrument as large as Fig. 168, on a fixed mount.

There are several rather elegant methods of adjusting the polar

[1] See also two valuable papers by Sir Howard Grubb, *The Observatory*, Vol. VII, pp. 9, 43. Also in Jour. Roy. Ast. Soc. Canada, Dec., 1921, Jan. 1922.

axis to the pole which depend on the use of special graticules in the eyepiece, or on auxiliary devices applied to the telescope, the general principle being automatically to provide for setting off the distance between Polaris and the pole at the proper hour angle. A beautifully simple one is that of Gerrish (*Pop. Ast.* **29**, 283.

The simple plan here outlined will generally, however, prove sufficient for ordinary purposes and where high precision is necessary one has to turn to the more conventional astronomical methods.

If one gives his telescope a permanent footing such as is shown in Fig. 171 adjustment has rarely to be repeated. With a pillar mount such as we have just now been considering the instrument itself can be taken in doors and replaced with very slight risk of disturbing its setting, but some provision must be made for sheltering the mount.

A tarpaulin is sometimes recommended and indeed answers well, particularly if a bag of rubber sheeting is drawn loosely over the mount first. Better still is a box cover of copper or galvanized iron set over the mount and closely fitting well down over a base clamped to the pillar with a gasket to close the joint.

But the fact is when one is dealing with a fine instrument like Fig. 168 of as much as 5 inches aperture, the question of a permanent housing (call it observatory if you like) at once comes up and will not down.

It is of course always more convenient to have the telescope permanently in place and ready for action. Some observers feel that working conditions are better with the telescope in the open, but most prefer a shelter from the wind, even if but partial, and the protection of a covering, however slight, in severe weather.

In the last resort the question is mainly one of climate. Where nights, otherwise of the best seeing quality, are generally windless or with breezes so slight that the tube does not quiver a telescope in the open, however protected between times, works perfectly well.

In other regions the clearest nights are apt to be those of a steady gentle wind producing great uniformity of conditions at the expense of occasional vibration of the instrument and of discomfort to the observer. Hence one finds all sorts of practice, varied too, by the inevitable question of expense.

The simplest possible housing is to provide for the fixed instrument a movable cover which can be lifted or slid quite out of the way leaving the telescope in the open air, exposed to wind, but free from the disturbing air currents that play around the opening of a dome. Shelters of this cheap and simple sort

FIG. 173.—The Simplest of Telescope Housings.

have been long in use both for small and large instruments.

For example several small astrographic instruments in the Harvard equipment are mounted as shown in Fig. 173. Here are two fork mounts, each on a short pier, and covered in by galvanized iron hoods made in two parts, a vertical door which swings down, as in the camera of the foreground, and the hood proper, hinged to the base plate and free to swing down when the rear door is unlocked and opened. A little to the rear is a similar astrographic camera with the hood closed. It is all very simple, cheap, and effective for an instrument not exceeding say two or three feet in focal length.

A very similar scheme has been successfully tried on reflectors as shown in Fig. 174. The instrument shown is a Browning equatorial of 8½ inches aperture. The cover is arranged to open after the manner of Fig. 173 and the plan proved very effective, preserving much greater uniformity of conditions and hence permitting better definition than in case of a similar instrument peering through the open shutter of a dome.

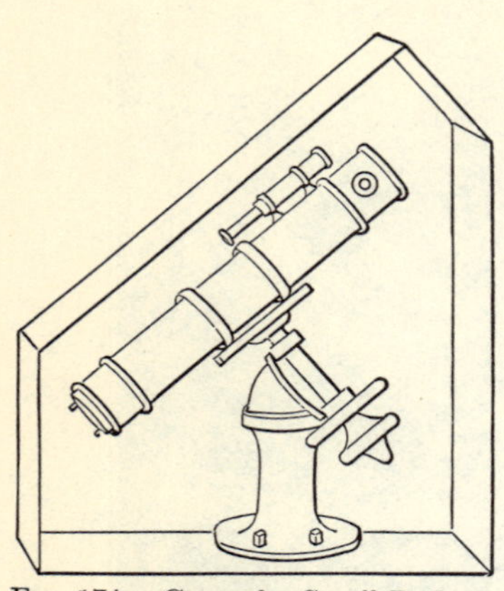

Fig. 174.—Cover for Small Reflector.

Such a contrivance gets unwieldly in case of a refractor on account of the more considerable height of the pier and the length of the tube itself. But a modification of it may be made to serve exceedingly well in climates where working in the open is advantageous. A good example is the equatorial of the Harvard Observatory station at Mandeville, Jamaica, which has been thus housed for some ten years, as shown in Fig. 175.

This 11-inch refractor, used mainly on planetary detail, is located alongside the polar telescope of 12 inches aperture and 135 feet 4 inches focal length used for making a photographic atlas of the moon and on other special problems. The housing, just big enough to take in the equatorial with the tube turned low, opens on the south side and then can be rolled northward on its track, into the position shown, where it is well clear of the instrument, which is then ready for use.

The climate of Jamaica, albeit extremely damp, affords remarkably good seeing during a large part of the year, and permits use of the telescope quite in the open without inconvenience to the observer. The success of this and all similar housing plans depends on the local climate more than on anything else—chiefly on wind during the hours of good seeing. An instrument quite uncovered suffers from gusts far more than one housed under a dome, which is really the sum of the whole matter, save that a dome to a slight extent does shelter the observer in extremely cold weather.

Even very large reflectors can be housed in similar fashion if suitably mounted. For example in Fig. 176 is shown the 36-inch aperture reflector of the late Dr. Common, which was fitted with an open fork equatorial mounting. Here the telescope itself, with its short pier and forked polar axis, is shown in dotted lines.

FIG. 175.—Sliding Housing for 11-inch Refractor.

Built about it is a combined housing and observing stand rotatable on wheels *T* about a circular track *R*. The housing consists of low corrugated metal sides and ends, here shown partly broken away, of dimensions just comfortably sufficient to take in the telescope when the housing is rotated to the north and south position, and the tube turned down nearly flat southward. A well braced track *WW* extends back along the top of the side housing and well to the rear. On this track rolls the roof of the housing *X*,*X*,*X*, with a shelter door at the front end.

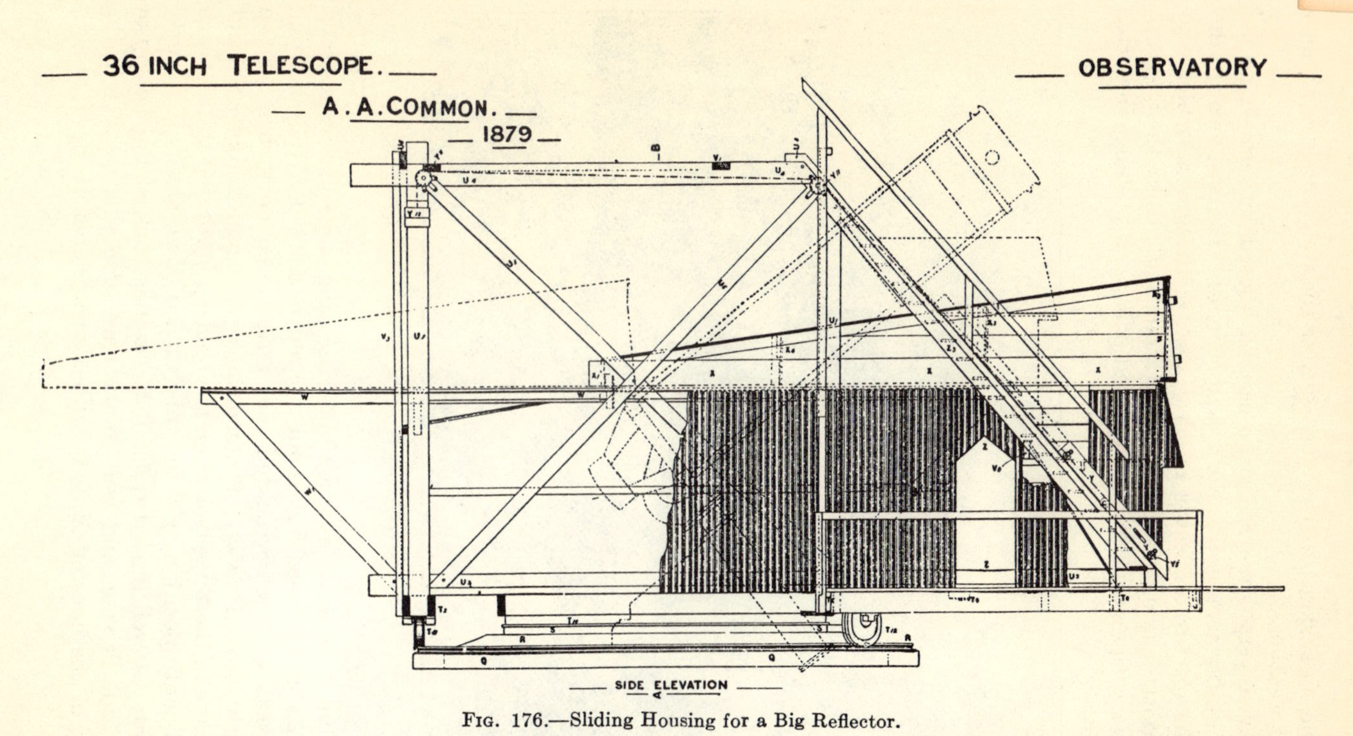

FIG. 176.—Sliding Housing for a Big Reflector.

The members *U* constitute a framing which supports at once the housing and the observing platform, to which access is had by a ladder, *Z*, provided with a counter-balanced observing seat. The instrument is put into action by clearing the door at the end of the roof, running the roof back to the position shown in the dotted lines, raising the tube, and then revolving the whole housing into whatever position is necessary to permit the proper setting of the tube.

FIG. 177.—Sliding Roof Observatory.

This arrangement worked well but was found a bit troublesome owing to wind and weather. With a skeleton tube and in a favorable climate the plan would succeed admirably providing an excellent shelter for a large telescope at very low cost.

Since a fork mount allows the tube to lie flat, such an instrument, up to say 8 or 10 inches aperture can be excellently protected by covers fitting snugly upon a base and light enough to lift off as a whole.

The successful use of all these shelters however depends on climatic conditions. They require circumstances allowing observation in the open, as with tripod mounts, and afford no protecion from wind or cold. Complete protection for the observer cannot be had, except by some of the devices shown in Chapter V, but conditions can be improved by permanent placement in an observatory, simple or elaborate, as the builder may wish.

The word observatory may sound formidable, but a modest one can be provided at less expense than a garage for the humblest motor car. The chief difference in the economic situation is that not even the most derided car can be picked up and carried into the back hall for shelter, and it really ought not to be left out in the weather.

The next stage of evolution is the telescope house with a sliding roof in one or more sections—ordinarily two. In this case the building itself is a simple square structure large enough to accommodate the instrument with maneuvering room around it. The side walls are carried merely high enough to give clearance to the tube when turned nearly flat and to give head room to the observer. The roof laps with a close joint in the middle and each half rolls on a track supported beyond the ends of the building by an out-rigger arranged in any convenient manner. When the telescope is in use the roof sections are displaced enough to give an ample clear space for observing, often wide open as shown in Fig. 177, which is the house of the 16-inch Metcalf photographic doublet at the Harvard Observatory. This instrument is in an open fork mount like that shown in Fig. 139.

The sliding roof type is on the whole the simplest structure that can be regarded as an observatory in the sense of giving some shelter to the observer as well as the instrument. It gives ample sky room for practical purposes even to an instrument with a fork mount, since in most localities the seeing within 30° or so of the horizon is decidedly bad. If view nearer the horizon is needed it can readily be secured by building up the pier a bit.

Numberless modifications of the sliding roof type will suggest themselves on a little study. One rather interesting one is used in the housing of the 24-inch reflector of the Harvard Observatory, 11 feet 3 inches in focal length, the same of which the drive in its original dome is shown in Fig. 139. As now arranged the lower part of the observatory remains while the upper works are quite similar in principle to the housing of Dr. Common's 3-foot reflector of Fig. 176. The cover open is shown in Fig. 178. It will be seen that on the north side of the observatory there is an outrigger on which the top housing slides clear of the low revolving turret which gives access to the ocular fitting used generally to carry the plate holder, and the eyepiece for following when required.

The tube cannot be brought to the horizontal, but it easily commands all the sky-space that can advantageously be used in this situation, and the protection given the telescope when not in use is very complete. To close the observatory the tube is brought north and south and turned low and the sliding roof is then run back into its fixed position. The turret is very easily turned by hand.

Fig. 178.—Turret Housing of the 24-inch Harvard Reflector.

Of course for steady work with the maximum shelter for observer obtainable without turning to highly special types of housing, the familiar dome is the astronomer's main reliance. It is in the larger sizes usually framed in steel and covered with wood, externally sheathed in copper or steel. Sometimes in smaller domes felt covered with rubberoid serves a good purpose, and painted canvas is now and then used, with wooden framing.

But even the smallest dome of conventional construction is heavy and rather expensive, and for home talent offers many difficulties, especially with respect to the shutter and shutter pening. A hemisphere is neither easy to frame nor to cover, nd the curved sliding shutter is especially troublesome.

Hence for small observatories other forms of revolving roof are desirable, and quite the easiest and cheapest contrivance is that embodied in the "Romsey" type of observatory, devised

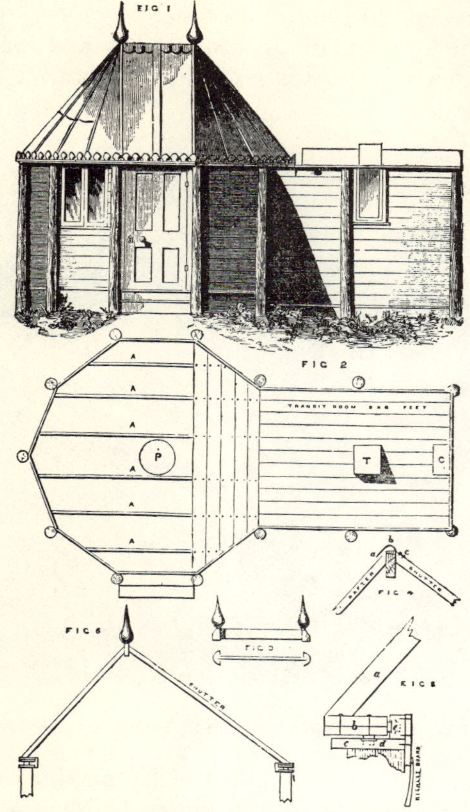

Fig. 179.—The Original "Romsey" Observatory.

half a century ago by that accomplished amateur the Rev. E. L. Berthon, vicar of Romsey. The feature of his construction is an unsymmetrical peak in the revolving roof which permits the ordinary shutter to be replaced by a hinged shutter like the skylight in a roof, exposing the sky beyond the zenith when open, and closing down over a coaming to form a water tight joint.

Berthon's original description of his observatory, which accommodated a 9¼-inch reflector, may be found in Vol. 14 of the *English Mechanic and World of Science* whence Fig. 179 is taken. In this plate Fig. 1 shows the complete elevation and Fig. 2 the ground plan, each to a scale of a eighth of an inch to the foot. In the plan, *A*,*A*, are the main joists, *P* the pier for the telescope, *T* that for the transit, and *C* the clock. Figs. 3, 4, and 5 are of details. In the last named *a* is a rafter, *b* the base ring, *c* the plate, *d* one of the sash rollers carrying the roof, and *e* a lateral guide roller holding the roof in place.

The structure can readily be built without the transit shelter, and in fact now-a-days most observers find it easier to pick up their time by wireless. The main bearing ring is cut out of ordinary ⅞ inch board, in ten or a dozen, or more, sections according to convenience, done in duplicate, joints lapping, and put very firmly together with screws set up hard. Sometimes 3 layers are thus used.

The roof in the original "Romsey" observatory was of painted canvas, but rubberoid or galvanized iron lined with roofing paper answers well. The shutter can be made single or double in width, and counterbalanced if necessary. The framing may be of posts set in the ground as here shown, or with sills resting on a foundation, and the walls of any construction—matched boards of any kind, cement on wire lath, hollow tile, or concrete blocks.

Chambers' *Handbook of Astronomy* Vol. II contains quite complete details of the "Romsey" type of observatory and is easier to get at than tne original description.

A very neat adaptation of the plan is shown in Fig. 180, of which a description may be found in *Popular Astronomy* **28,** 183. This observatory was about 9 feet in diameter, to house a 4-inch telescope, and was provided with a rough concrete foundation on which was built a circular wall 6 feet high of hollow glazed tile, well levelled on top. To this was secured a ring plate built up in two layers, carrying two circles of wooden strips with a couple of inches space between them for a run-way. In this ran 6 two-inch truck castors secured to a similar ring plate on which was built up the frame of the " dome" arranged as shown. Altogether a very neat and workmanlike affair, in this case built largely by the owner but permitting construction at very small expense almost anywhere. Another interesting modification of the same general plan in the same volume just cited is shown in Fig. 181. This is also for a 4-inch refractor and the dome proper is but 8 feet 4 inches in diameter. Like the preceding structure the foundation is of concrete but the walls are framed in spruce and sheathed in matched boards with a "beaver-board" lining.

FIG. 180.—A More Substantial "Romsey" Type.

The ring plate is three-ply, 12 sections to the layer, and its mate on which the dome is assembled is similarly formed, though left with the figure of a dodecagon to match the dome. The weight is carried on four rubber tired truck rollers, and there are lateral guide rollers on the plan of those in Fig. 179.

The dome itself however, is wholly of galvanized iron, in 12 gores joined with standing seams, turned, riveted, and soldered.

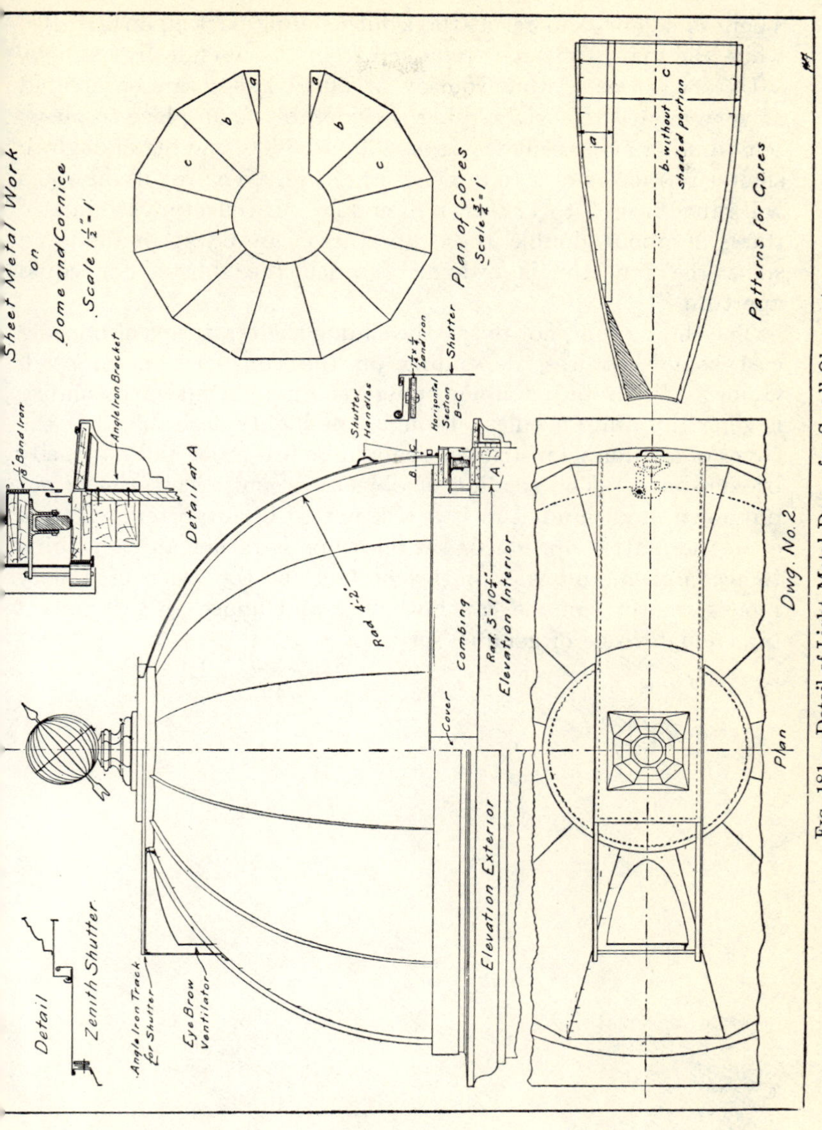

FIG. 181.—Detail of Light Metal Dome for Small Observatory.

There is a short shutter at the zenith sliding back upon a frame, while the main shutter is removed from the outside by handles.

Observatories of the Romsey or allied types can be erected at very moderate cost, varying considerably from place to place, but running at present say from $200 to $600, and big enough to shelter refractors of 4 to 6 inches aperture. The revolving roofs will range from 9 to 12 feet in diameter. If reflectors are in use, those of about double these apertures can be accommodated since the reflector is ordinarily much the shorter for equal aperture.

The sliding roof, not to say the sliding shelter, forms of housing cost somewhat less, depending on the construction adopted. Going to brick may double the figures quoted, but such solidity is generally quite needless, though it is highly desirable that the cover of a valuable instrument should be fire-proof and not easily broken open. The stealing of objectives and accessories is not unknown, and vandalism is a risk not to be forgotten. But to even the matter up, housing a telescope is rather an easy thing to accomplish, and as a matter of fact for the price of a very modest motor car one can both buy and house an instrument big enough to be of genuine service.

CHAPTER XI

SEEING AND MAGNIFICATION

Few things are more generally disappointing than one's first glimpse of the Heavens through a telescope. The novice is fed up with maps of Mars as a great disc full of intricate markings, and he generally sees a little wriggling ball of light with no more visible detail than an egg. It is almost impossible to believe that, at a fair opposition, Mars under the power of even the smallest astronomical telescope really looks as big as the full moon. Again, one looks at a double star to see not two brilliant little discs resplendent in color, but an indeterminate flicker void of shape and hue.

The fact is, that most of the time over most of the world seeing conditions are bad, so that the telescope does not have a fair chance, and on the whole the bigger the telescope the worse the chance. One famous English astronomer, possessed of a fine refractor that would be reckoned large even now-a-days, averred that he had seen but one first class night in fifteen years past.

The case is really much less bad than this implies, for even in rather unfavorable climates many a night, at some o'clock or other, will furnish an hour or two of pretty good seeing, while now and then, without any apparent connection with the previous state of the weather, a night will turn up when the pictures in the popular astronomies come true, the stars shrink to steady points set in clean cut rings, and no available power seems too high.

One can get a good idea of the true inwardness of bad seeing by trying to read a newspaper through an opera glass across a hot stove. If the actual movements in the atmosphere could be made visible they would present a strange scene of turbulence—rushing currents taking devious courses up and around obstacles, slowly moving whirlpools, upward slants such as gulls hug on the quarter of a liner, great downward rushes dreaded by the aviator, and over it all incessant ripples in every direction.

And movements of air are usually associated with changes of temperature, as over the stove, varying the refraction and contorting the rays that come from a distant star until the image is quite ruined.

The condition for excellence of definition is that the atmosphere through which we see shall be homogeneous, whatever its temperature, humidity, or general trend of movement. Irregular refraction is the thing to be feared, particularly if the variations are sudden and frequent. Hence the common troubles near the ground and about buildings, especially where there are roofs and chimneys to radiate heat—even in and about an observatory dome.

Professor W. H. Pickering, who has had a varied experience in climatic idiosyncrasies, gives the Northern Atlantic seaboard the bad preëminence of having the worst observing conditions of any region within his knowledge. The author cheerfully concurs, yet now and then, quite often after midnight, the air steadies and, if the other conditions are good, definition becomes fairly respectable, sometimes even excellent.

Temperature and humidity as such, seem to make little difference, and a steady breeze unless it shakes the instrument is relatively harmless. Hence we find the most admirable definition in situations as widely different as the Harvard station at Mandeville, Jamaica; Flagstaff, Arizona 7000 feet up and snow bound in winter; Italy, and Egypt. The first named is warm and with very heavy rainfall and dew, the second dry with rather large seasonal variation of temperature, and the others temperate and hot respectively.

Perhaps the most striking evidence of the importance of uniformity was noted by Evershed at an Indian station where good conditions immediately followed the flooding of the rice fields with its tendency to stabilize the temperature. Mountain stations may be good as at Flagstaff, Mt. Hamilton, or Mt. Wilson, or very bad as Pike's Peak proved to be, probably owing to local conditions.

In fact much of the trouble comes from nearby sources, atmospheric waves and ripples rather than large movements, ripples indeed often small compared with the aperture of the telescope and sometimes in or not far outside of the tube itself.

Aside from these difficulties, there are still others which have to do with the transparency of the atmosphere with respect to its suspended matter. This does not affect the definition as such, but it cuts down the light to a degree that may interfere seriously with the observation of faint stars and nebulae. The smoke near a city aggravates the situation, but in particular it depends on general weather conditions which may be persistent or merely temporary.

Often seeing conditions may be admirable save for this lack of transparency in the atmosphere, so that study of the moon, of planetary markings and even of double stars, not too faint, may go on quite unimpeded. The actual loss of light may reach however a magnitude or more, while the sky is quite cloudless and without a trace of fog or noticeable haziness by day.

There have been a good many nights the past year (1921) when Alcor (80 Ursae Majoris) the tiny neighbor of Mizar, very nearly of the 4th magnitude, has been barely or not at all visible while the seeing otherwise was respectably good. Ordinarily stars of 6^m should be visible in a really clear night, and in a brilliant winter sky in the temperate zones, or in the clear air of the tropics, a good many eyes will do better than this, reaching $6^m.5$ or even 7^m, occasionally a bit more.

The relation of air waves and such like irregularities to telescopic vision was rather thoroughly investigated by Douglass more than twenty years ago (Pop. Ast. **6,** 193) with very interesting results. In substance, from careful observation with telescopes from 4 inches up to 24 inches aperture, he found that the real trouble came from what one may call ripples, disturbances from say 4 inches wave length down to ¾ inch or less. Long waves are rare and relatively unimportant since their general effect is to cause shifting of the image as a whole rather than the destruction of detail which accompanies the shorter waves.

This rippling of the air is probably associated with the contact displacements in air currents such as on a big scale become visible in cloud forms. Clearly ripples, marked as they are by difference of refraction, located in front of a telescope objective, produce different focal lengths for different parts of the objective and render a clean and stable image quite out of the question.

In rough terms Douglass found that waves of greater length than half the aperture did not materially deteriorate the image, although they did shift it as a whole, while waves of length less than one third the aperture did serious mischief to the definition, the greater as the ripples were shorter, and the image itself more minute in dimension or detail.

Hence there are times when decreasing the aperture of an objective by a stop improves the seeing considerably by increasing the relative length of the air waves. Such is in fact found to be the case in practical observing, especially when the seeing with a large aperture is decidedly poor. In other words one may often gain more by increased steadiness than he loses by lessened "resolving power," the result depending somewhat on the class of observation which chances to be under way.

And this brings us, willy-nilly, to the somewhat abstruse matter of resolving power, depending fundamentally upon the theory of diffraction of light, and practically upon a good many other things that modify the character of the diffraction pattern, or the actual visibility of its elements.

When light shines through a hole or a slit the light waves are bent at the margins and the several sets, eventually overlapping, interfere with each other so as to produce a pattern of bright and dark elements depending on the size and shape of the aperture, and distributed about a central bright image of that aperture. One gets the effect well in looking through an open umbrella at a distant street light. The outer images of the pattern are fainter and fainter as they get away from the central image.

Without burdening the reader for the moment with details to be considered presently, the effect in telescopic vision is that a star of real angular diameter quite negligible, perhaps 0.″001 of arc, is represented by an image under perfect conditions like Fig. 154, of quite perceptible diameter, surrounded by a system of rings, faint but clear-cut, diminishing in intensity outwards. When the seeing is bad no rings are visible and the central disc is a mere bright blur several times larger than it ought to be.

The varying appearance of the star image is a very good index of the quality of the seeing, so that, having a clear indication of this appearance, two astronomers in different parts of the world can gain a definite idea of each other's relative seeing conditions. To this end a standard scale of seeing, due largely to the efforts of Prof. W. H. Pickering, has come into rather common use. (H. A. **61,** 29). It is as follows, based on observations with a 5-inch telescope.

STANDARD SCALE OF SEEING

1. Image usually about twice the diameter of the third ring.
2. Image occasionally twice the diameter of the third ring.

3. Image of about the same diameter as the third ring, and brighter at the centre.

4. Disc often visible, arcs (of rings) sometimes seen on brighter stars.

5. Disc always visible, arcs frequently seen on brighter stars.

6. Disc always visible, short arcs constantly seen.

7. Disc sometimes sharply defined, (*a*) long arcs. (*b*) Rings complete.

8. Disc always sharply defined, (*a*) long arcs. (*b*) Rings complete all in motion.

9. Rings. (*a*) Inner ring stationary, (*b*) Outer rings momentarily stationary.

10. Rings all stationary. (*a*) Detail between the rings sometimes moving. (*b*) No detail between the rings.

The first three scale numbers indicate very bad seeing; the next two, poor; the next two, good; and the last three, excellent. One can get some idea of the extreme badness of scale divisions 1, 2, 3, in realizing that the third bright diffraction ring is nearly 4 times the diameter of the proper star-disc.

It must be noted that for a given condition of atmosphere the seeing with a large instrument ranks lower on the scale than with a small one, since as already explained the usual air ripples are of dimensions that might affect a 5 inch aperture imperceptibly and a 15-inch aperture very seriously.

Douglass (loc. cit.) made a careful comparison of seeing conditions for apertures up to 24 inches and found a systematic difference of 2 or 3 scale numbers between 4 or 6 inches aperture, and 18 or 24 inches. With the smallest aperture the image showed merely bodily motion due to air waves that produced serious injury to the image in the large apertures, as might be expected.

There is likewise a great difference in the average quality of seeing as between stars near the zenith and those toward the horizon, due again to the greater opportunity for atmospheric disturbances in the latter case. Pickering's experiments (loc. cit.) show a difference of nearly 3 scale divisions between say 20° and 70° elevation. This difference, which is important, is well shown in Fig. 182, taken from his report.

The three lower curves were from Cambridge observations, the others obtained at various Jamaica stations. They clearly show the systematic regional differences, as well as the rapid falling off in definition below altitude 40°, which points the importance of making provision for comfortable observing above this altitude.

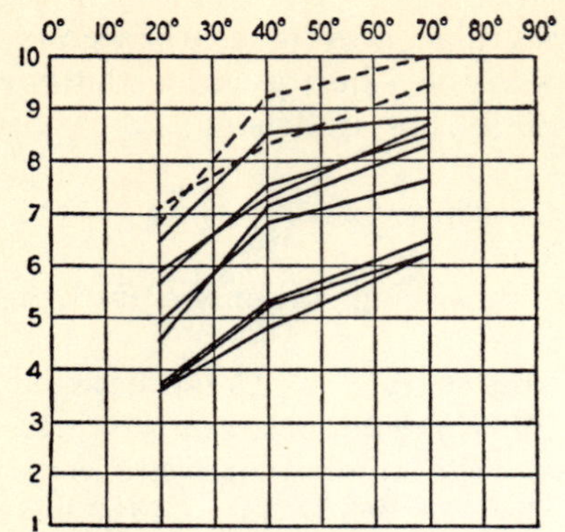

FIG. 182.—Variation of Seeing with Altitude.

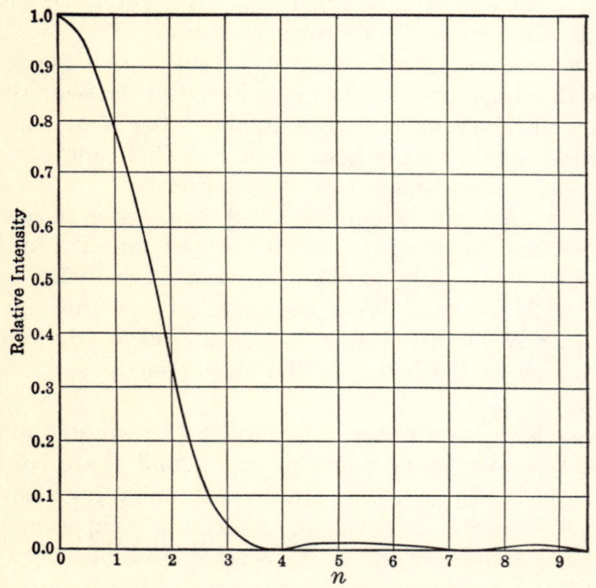

FIG. 183.—Airy's Diffraction Pattern.

The relation of the diffraction pattern as disclosed in the moments of best seeing to its theoretical form is a very interesting one. The diffraction through a theoretically perfect objective was worked out many years ago by Sir George Airy who calculated the exact distribution of the light in the central disc and the surrounding rings.

This is shown from the centre outwards in Fig. 183, in which the ordinates of the curve represent relative intensities while the abscissæ represent to an arbitrary scale the distances from the axis. It will be at once noticed that the star image, brilliant at

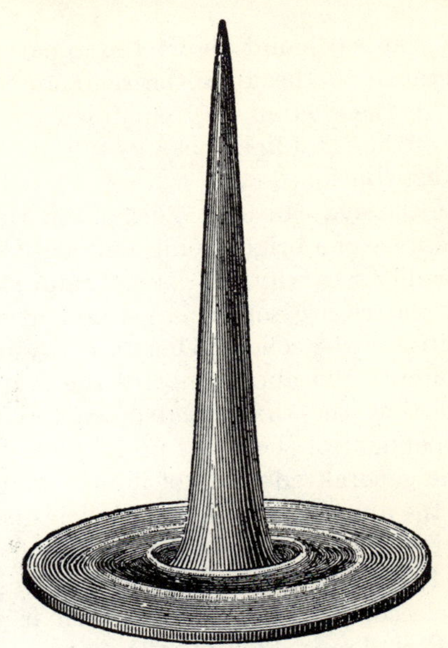

FIG. 184.—Diffraction Solid for a Star.

its centre, sinks, first rapidly and then more slowly, to a minimum and then very gradually rises to the maximum of the first bright ring, then as slowly sinks again to increase for the second ring and so on.

For unity brightness in the centre of the star disc the maximum brightness of the first ring is 0.017, of the second 0.004 and the third 0.0016. The rings are equidistant and the star disc has a radius substantially equal to the distance between rings. One's vision does not follow down to zero the intensities of the rings or of the margin of the disc, so that the latter has an apparent diameter materially less than the diameter to the first diffraction minimum, and the rings themselves look sharper and thinner than the figure would show, even were the horizontal scale much diminished. The eye does not descend in the presence of bright areas to its final threshold of perception.

One gains a somewhat vivid idea of the situation by passing to three dimensions as in Fig. 184, the "diffraction solid" for a star, a conception due to M. André (Mem. de l'Acad. de Lyon **30,** 49). Here the solid represents in volume the whole light received and the height taken at any point, the intensity at that point.

A cross section at any point shows the apparent diameter of the disc, its distance to the apex the remaining intensity, and the volume above the section the remaining total light. Substantially 85 % of the total light belongs to the central cone, for the theoretical distribution.

Granting that the eye can distinguish from the back-ground of the sky, in presence of a bright point, only light above a certain intensity, one readily sees why the discs of faint stars look small, and why shade glasses are sometimes useful in wiping out the marginal intensities of the solid. There are physiological factors that alter profoundly the appearance of the actual star image, despite the fact that the theoretical diffraction image for the aperture is independent of the star's magnitude.

Practically the general reduction of illumination in the fainter stars cuts down the apparent diameters of their discs, and reduces the number of rings visible against the background of the sky.

The scale of the diffraction system determines the resolving power of the telescope. This scale is given in Airy's original paper (Cambr. Phil. Trans. **1834** p. 283), from which the angle α to any maximum or minimum in the ring system is defined by

$$\sin \alpha = n \frac{\lambda}{R}$$

in which λ is numerically the wave length of any light considered and R is the radius of the objective.

We therefore see that the ring system varies in dimension inversely with the aperture of the objective and directly with the wave length considered. Hence the bigger the objective the smaller the disc and its surrounding ring system; and the greater the wave length, i.e. the redder the light, the bigger the diffraction system. Evidently there should be color in the rings but it very seldom shows on account of the faintness of the illumination.

Now the factor n is for the first dark ring 0.61, and for the first bright ring 0.81, as computed from Airy's general theory, and therefore if we reckon that two stars will be seen as separate when the central disc of one falls on the first dark ring of the other the angular distance will be

$$\text{Sin}\,\alpha = 0.61\,\frac{\lambda}{R},$$

and, taking λ at the brightest part of the spectrum i.e., about 560 $\mu\mu$, in the yellow green, with α taken for sin α, we can compute this assumed separating power for any aperture. Thus 560 $\mu\mu$ being very nearly $1/_{45,500}$ inch, and assuming a 5-inch telescope, the instrument should on this basis show as double two stars whose centres are separated by 1.″1 of arc.

In actual fact one can do somewhat better than this, showing that the visible diameter of the central disc is in effect less than the diameter indicated by the diffraction pattern, owing to the reasons already stated. Evidently the brightness of the star is a factor in the situation since if very bright the disc gains apparent size, and when very faint there is sufficient difficulty in seeing one star, let alone a pair.

The most thorough investigation of this matter of resolving power was made by the Rev. W. R. Dawes many years ago (Mem. R.A.S. **35,** 158). His study included years of observation with telescopes of different sizes, and his final result was to establish what has since been known as "Dawes' Limit."

To sum up Dawes' results he established the fact that on the average a one-inch aperture would enable one to separate two 6th magnitude stars the centers of which were separated by 4″.56. Or, to generalize from this basis, the separating power of any telescope is for very nearly equal stars, moderately bright, $\frac{4''.56}{A}$ where A is the aperture of the telescope in inches.

Many years of experience have emphasized the usefulness of this approximate rule, but that it is only approximate must be candidly admitted. It is a limit decidedly under that just assigned on the basis of the theory of diffraction for the central bright wave-lengths of the spectrum. Attempts have been made to square the two figures by assuming in the diffraction theory a wave length of $1/_{55,000}$ inch, but this figure corresponds to a point well up into the blue, of so low luminosity that it is of no importance whatever in the visual use of a telescope.

The fact is that the visibility of two neighboring bright points as distinct, depends on a complex of physical and physiological factors, the exact relations of which have never been unravelled. To start with we have the principles of diffraction as just explained, which define the relation of the stellar disc to the center of the first dark ring, but we know that under no circumstances can one see the disc out to this limit, since vision fails to take cognizance of the faint rim of the image. The apparent diameter of the diffraction solid therefore corresponds to a section taken some distance above the base, the exact point depending on the sensitiveness of the particular observer's eye, the actual brilliancy of the center of the disc, and the corresponding factors for the neighboring star.

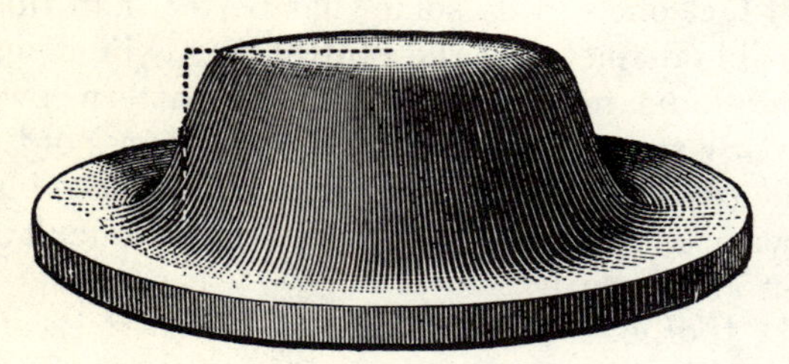

Fig. 185.—Diffraction Solid for a Disc.

Under favorable circumstances one would not go far amiss in taking the visible diameter of the disc at about half that reckoned to the center of the first dark ring. This figure in fact corresponds to what has been shown to be within the grasp of a good observer under favorable conditions, as we shall presently see.

On the other hand, if the stars are decidedly bright there is increase of apparent diameter of the disc due to the phenomenon known as irradiation, the spreading of light about its true image on the retina which corresponds quite closely to the halation produced by a bright spot on a photographic plate.

If, on the contrary, the stars are very faint the total amount of light available is not sufficient to make contrast over and above the background sufficient to disclose the two points as separate, while if the pair is very unequal the brighter one will produce sufficient glare to quite over-power the light from the smaller one so that the eye misses it entirely.

A striking case of this is found in the companion to Sirius, an extremely difficult object for ordinary telescopes although the distance to the companion is about $10''.6$ and its magnitude is 8.4, making a superlatively easy double for the very smallest telescope save for the overpowering effect of the light of the large star. Another notoriously difficult object for small telescopes is δ Cygni, a beautiful double of which the smaller component falls unpleasantly near the first diffraction maximum of the primary in which it is apt to be lost.

"Dawes' Limit" is therefore subject to many qualifying factors. Lewis, in the papers already referred to (Obs. **37,** 378) did an admirable piece of investigation in going through the double star work of about two score trained observers working with telescopes all the way from 4 inches to 36 inches aperture.

From this accumulation of data several striking facts stand out. First there is great difference between individual observers working with telescopes of similar aperture as respects their agreement with "Dawes' Limit," showing the effect of variation in the physiological factors as well as instrumental ones.

Second, there is also a very large difference between the facility of observing equal bright pairs and equal faint pairs, or unequal pairs of any kind, again emphasizing the physiological as well as the physical factors.

Finally, there is most unmistakable difference between small and large apertures in their capacity to work up to or past the standard of "Dawes' Limit." The smaller telescopes are clearly the more efficient as would be anticipated from the facts just pointed out regarding the different effect of the ordinary and inescapable atmospheric waves on small and large instruments.

The big telescopes are unquestionably as good optically speaking as the small ones but under the ordinary working conditions, even as good as those a double star observer seeks, the smaller aperture by reason of less disturbance from atmospheric factors does relatively much the better work, however good the big instrument may be under exceptional conditions.

This is admirably shown by the discussion of the beautiful work of the late Mr. Burnham, than whom probably no better observer of doubles has been known to astronomy. His records of discovery with telescopes of 6, 9.4, 12, 18½ and 36 inches show the relative ease of working to the theoretical limit with instruments not seriously upset by ordinary atmospheric waves.

With the 6-inch aperture Burnham reached in the average 0.53 of Dawes' limit, quite near the rough figure just suggested, and he also fell well inside Dawes' limit with the 9.4-inch instrument.

With none of the others did he reach it and in fact fell short of it by 15 to 60%. All observations being by the same notably skilled observer and representing discoveries of doubles, so that no aid could have been gained by familiarity, the issue becomes exceedingly plain that size with all its advantages in resolving power brings serious countervailing limitations due to atmosphere.

But a large aperture has besides its possible separating power one advantage that can not be discounted in "light grasp," the power of discerning faint objects. This is the thing in which a small telescope necessarily fails. The "light grasp" of the telescope obviously depends chiefly on the area of the objective, and visually only in very minor degree on the absorption of the thicker glass in the case of a large lens.

According to the conventional scale of star magnitudes as now in universal use, stars are classified in magnitudes which differ from each other by a light ratio of 2.512, a number the logarithm of which is 0.4, a relation suggested by Pogson some forty years ago. A second magnitude star therefore gives only about 40% of the light of a first magnitude star, while a third magnitude star gives again a little less than 40% of the light of a second magnitude star and so on.

But doubling the aperture of a telescope increases the available area of the objective four times and so on, the "light grasp" being in proportion to the square of the aperture. Thus a 10-inch objective will take in and deliver nearly 100 times as much light as would a 1-inch aperture. If one follows Pogson's scale down the line he will find that this corresponds exactly to 5 stellar magnitudes, so that if a 1-inch aperture discloses, as it readily does, a 9th magnitude star, a 10-inch aperture should disclose a 14th magnitude star.

Such is substantially in fact the case, and one can therefore readily tabulate the minimum visible for an aperture just as he can tabulate the approximate resolving power by reference to Dawes' limit. Fig. 186 shows in graphic form both these relations for ready reference, the variation of resolving power with aperture, and that of "light grasp," reckoned in stellar magnitudes.

It is hardly necessary to state that considerable individual and observational differences will be found in each of these cases, in the latter amounting to not less than 0.5 to 1.0 magnitude either way. The scale is based on the 9th magnitude star just being visible with 1-inch aperture, whereas in fact under varying conditions and with various observers the range may be from the 8th to 10th magnitude. All these things, however convenient, must be taken merely at their true value as good working approximations.

Even the diffraction theory can be taken only as an approximation since no optical surface is absolutely perfect and in the ordinary refracting telescope there is a necessary residual chromatic aberration beside whatever may remain of spherical errors.

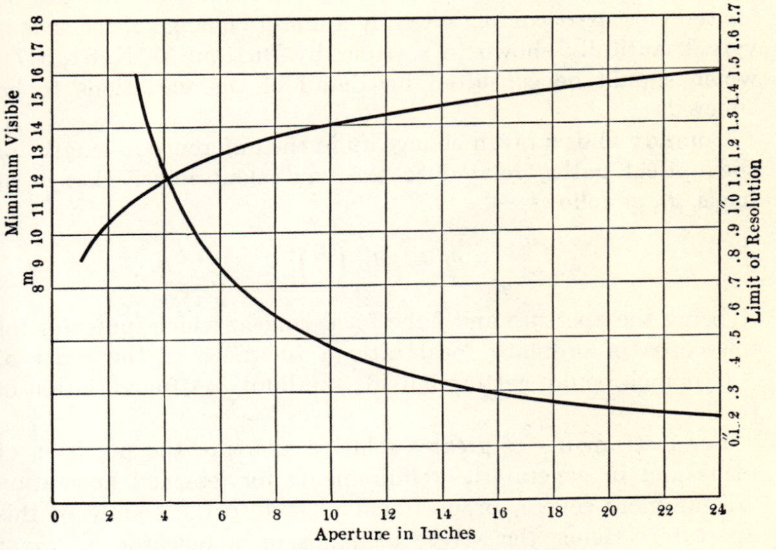

FIG. 186.—Light-grasp and Resolving Power.

It is a fact therefore, as has been shown by Conrady (M.N. **79**, 575) following up a distinguished investigation by Lord Rayleigh (Sci. Papers **1,** 415), that a certain small amount of aberration can be tolerated without material effect on the definition, which is very fortunate considering that the secondary spectrum represents aberrations of about $1/2{,}000$ of the focal length, as we have already seen.

The chief effect of this, as of very slight spherical aberration, is merely to reduce the maximum intensity of the central disc of the diffraction pattern and to produce a faint haze about it which slightly illuminates the diffraction minima. The visible diameter of the disc and the relative distribution of intensity in it is not however materially changed so that the main effect is a little loss and scattering of light.

With larger aberrations these effects are more serious but where the change in length of optical path between the ray proceeding through the center of the objective and that from the margin does not exceed $\frac{1}{4}\lambda$ the injury to the definition is substantially negligible and virtually disappears when the image is focussed for the best definition, the loss of maximum intensity in the star disc amounting to less than 20%.

Even twice this error is not a very serious matter and can be for the most part compensated by a minute change of focus as is very beautifully shown in a paper by Buxton(M. N. **81,** 547), which should be consulted for detail of the variations to be effected.

Conrady finds a given change dp in the difference in lengths of the optical paths, related to the equivalent linear change of focus, df, as follows:—

$$df = 8dp \left(\frac{f}{A}\right)^2$$

A being the aperture and f the focal length, which indicates for telescopes of ordinary focal ratio a tolerance of the order of ± 0.01 inch before getting outside the limit $\frac{1}{4}\lambda$ for variation of path.

For instruments of greater relative aperture the precision of focus and in general the requirements for lessened aberration are far more severe, proportional in fact to the square of this aperture. Hence the severe demands on a reflector for exact figure. An instrument working at F/5 or F/6 is extremely sensitive to focus and demands great precision of figure to fall within permissible values, say $\frac{1}{4}\lambda$ to $\frac{1}{2}\lambda$, for dp.

Further, with a given value of dp and the relation established by the chromatic aberration, *i.e.*, about $\frac{f}{2000}$, a relation is also determined between f and A, required to bring the aberration within limits. The equation thus found is

$$f = 2.8A^2$$

This practically amounts to the common F/15 ratio for an aperture of approximately 5 inches. For smaller apertures a greater ratio can be well used, for larger, a relatively longer focus is indicated, the penalty being light spread into a halo over the diffraction image and reducing faint contrasts somewhat seriously.

This is one of the factors aside from atmosphere, interfering with the full advantage of large apertures in refractors. While as already noted small amounts of spherical aberration may be to a certain extent focussed out, the sign of *df* must change with the sign of the residual aberration, and a quick and certain test of the presence of spherical aberration is a variation in the appearance of the image inside and outside focus.

To emphasize the importance of exact knowledge of existing aberrations note Fig. 187, which shows the results of Hartmann tests on a typical group of the world's large objectives. All show traces of residual zones, but differing greatly in magnitude and position as the attached scales show. The most conspicuous aberrations are in the big Potsdam photographic refractor, the least are in the 24-inch Lowell refractor. The former has since been refigured by Schmidt and revised data are not yet available; the latter received its final figure from the Lundins after the last of the Clarks had passed on.

Now a glance at the curves shows that the bad zone of the Potsdam glass was originally near the periphery, (I), hence both involved large area and, from Conrady's equation, seriously enlarged *df* due to the large relative aperture at the zone. An aberrant zone near the axis as in the stage (III) of the Potsdam objective or in the Ottawa 15-inch objective is much less harmful for corresponding reasons. Such differences have a direct bearing on the use of stops, since these may do good in case of peripheral aberration and harm when the faults are axial. Unless the aberrations are known no general conclusions can be drawn as to the effect of stops. Even in the Lowell telescope shown as a whole in Fig. 188, the late Dr. Lowell found stops to be useful in keeping down atmospheric troubles and reducing the illumination although they could have had no effect in relation to figure. Fig. 188 shows at the head of the tube a fitting for a big iris diaphragm, controlled from the eye-end, the value of which was well demonstrated by numerous observers.

There are, too, cases in which a small instrument, despite intrinsic lack of resolving power, may actually do better work than a big one. Such are met in instances where extreme con-

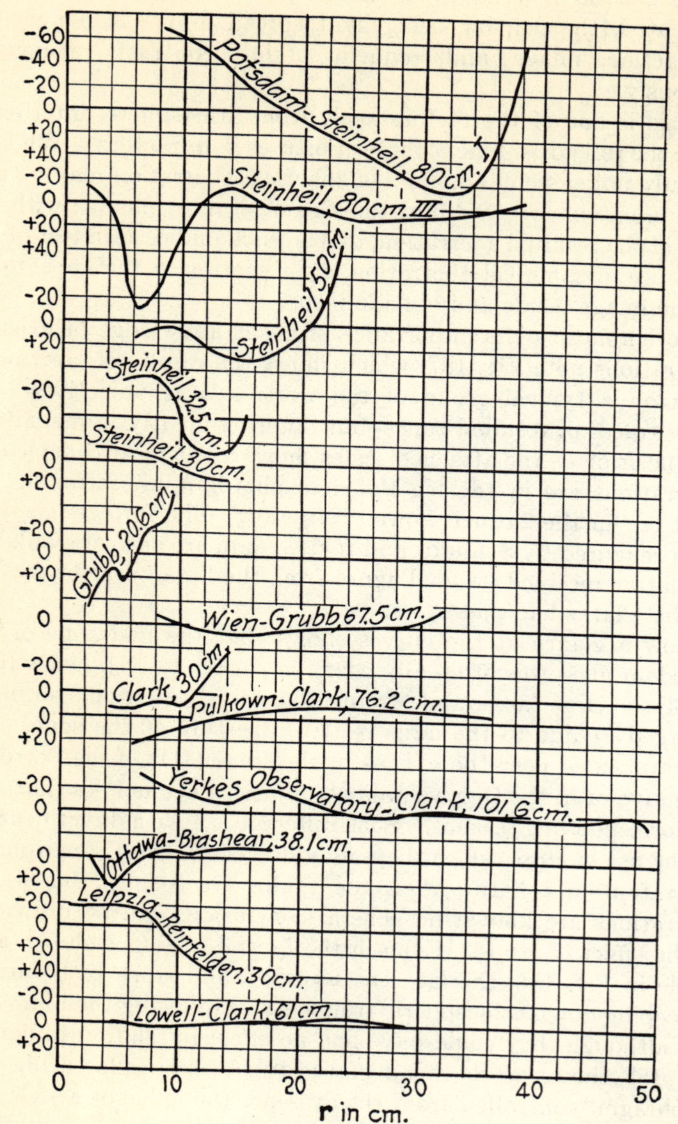

FIG. 187.—Hartmann Tests of Telescopes [From Hartmann's Measures].

trast of details is sought, as has been well pointed out by Nutting (Ap. J. **40,** 33) and the situation disclosed by him finds amplification in the extraordinary work done by Barnard with a cheap lantern lens of 1½ inch diameter and 5½ inches focus (Pop. Ast., **6,** 452).

The fact is that every task must seek its own proper instrument. And in any case the interpretation of observed results is a matter that passes far beyond the bounds of geometrical optics, and involves physiological factors that are dominant in all visual problems.

With respect to the visibility of objects the general diffraction theory again comes into play. For a bright line, for example, the diffraction figure is no longer chiefly a cone like Fig. 183, but a similar long wedge-shaped figure, with wave-like shoulders corresponding to the diffraction rings. The visibility of such a line depends not only on the distribution of intensity in the theoretical wedge but on the sensitiveness of the eye and the nature of the background and so forth, just as in the case of a star disc.

If the eye is from its nature or state of adaptation keen enough on detail but not particularly sensitive to slight differences of intensity, the line will very likely be seen as if a section were made of the wedge near its thin edge. In other words the line will appear thin and sharp as the diffraction rings about a star frequently do.

With an eye very sensitive to light and small differences of contrast the appearance of absolutely the same thing may correspond to a section through the wedge near its base, in other words to a broad strip shading off somewhat indistinctly at the edges, influenced again by irradiation and the character of the background.

If there be much detail simultaneously visible the diffraction patterns may be mixed up in a most intricate fashion and one can readily see the confusion which may exist in correlating the work of various observers on things like planetary and lunar detail.

In the planetary case the total image is a complex of illuminated areas of diffraction at the edges, which may be represented as the diffraction solid of Fig. 185, in which the dotted lines show what may correspond fairly to the real diameter of the planet, the edge shading off in a way again complicated by irradiation.

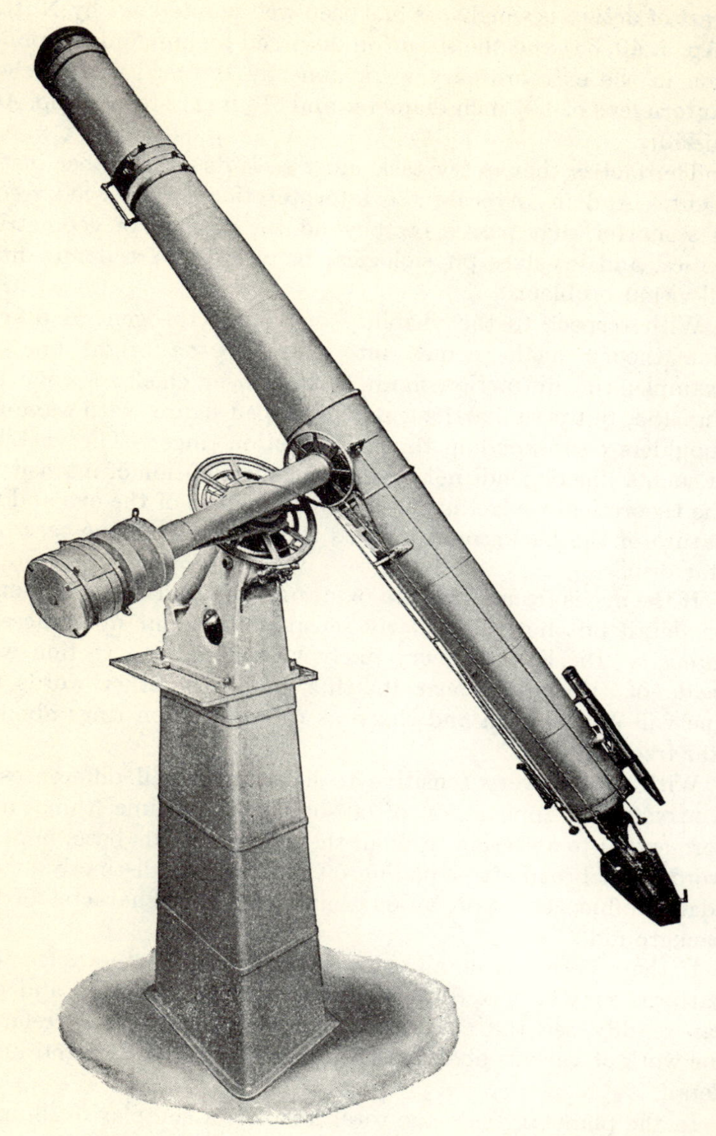

FIG. 188.—The Lowell Refractor Fitted with Iris Diaphragm.

Fancy detail superimposed on a disc of this sort and one has a vivid idea of the difficulty of interpreting observations.

It would be an exceedingly good thing if everyone who uses his telescope had the advantage of at least a brief course in microscopy, whereby he would gain very much in the practical understanding of resolving power, seeing conditions, and the interpretation of the image. The principles regarding these matters are in fact very much the same with the two great instruments of research.

Aperture, linear in the case of the telescope and the so-called numerical in the case of the microscope, bear precisely the same relation to resolution, the minimum resolvable detail being in each case directly proportional to aperture in the senses here employed.

Further, although the turbulence of intervening atmosphere does not interfere with the visibility of microscopic detail, a similar disturbing factor does enter in the form of irregular and misplaced illumination. It is a perfectly easy matter to make beautifully distinct detail quite vanish from a microscopic image merely by mismanagement of the illumination, just as unsteady atmosphere will produce substantially the same effect in the telescopic image.

In the matter of magnification the two cases run quite parallel, and magnification pushed beyond what is justified by the resolving power of the instrument does substantially little or no good. It neither discloses new detail nor does it bring out more sharply detail which can be seen at all with a lower power.

The microscopist early learns to shun high power oculars, both from their being less comfortable to work with, and from their failing to add to the efficiency of the instrument except in some rare cases with objectives of very high resolving power. Furthermore in the interpretation of detail the lessons to be learned from the two instruments are quite the same, although one belongs to the infinitely little and the other to the infinitely great.

Nothing is more instructive in grasping the relation between resolving power, magnification, and the verity of detail, than the study under the microscope of some well known objects. For example, in Fig. 189 is shown a rough sketch of a common diatom, *Navicula Lyra*. The tiny siliceous valve appears thus under an objective of slightly insufficient resolving power. The general form of the object is clearly perceived, as well as the central markings, standing boldly out in the form which suggests the specific name. No trace of any finer detail appears and no amount of dexterity in arranging the illumination or increase of magnifying power will show any more than here appears, the drawing beng one actually made with the camera lucida, using an objective of numerical aperture just too small to resolve the details of the diatoms on this particular slide.

Figure 189*a* shows what happens when, with the same magnifying power, an objective of slightly greater aperture is employed. Here the whole surface of the valve is marked with fine stria-

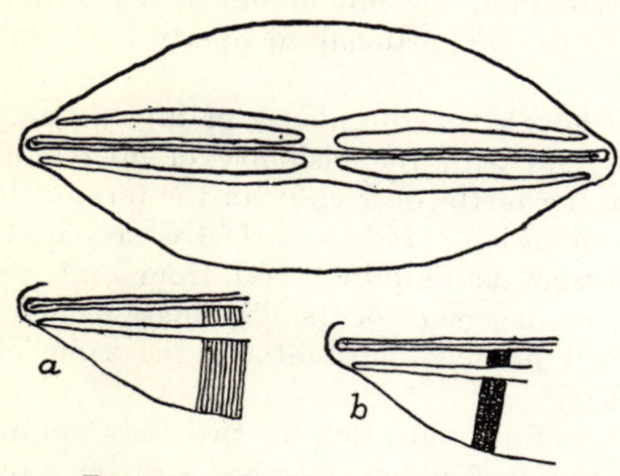

Fig. 189.—The Stages of Resolution.

tions, beautifully sharp and distinct like the lines of a steel engraving. There is a complete change of aspect wrought by an increase of about 20% in the resolving power. Again nothing further can be made out by an increase of magnification, the only effect being to make the outlines a little hazier and the view therefore somewhat less satisfactory.

Finally in Fig. 189*b* we have again the same valve under the same magnifying power, but here obtained from an objective of numerical aperture 60% above that used for the main figure. The sharp striæ now show their true character. They had their origin in lines of very clearly distinguished dots, which are perfectly distinct, and are due to the resolving power at last being sufficient to show the detail which previously merely formed a sharp linear diffraction pattern entirely incapable of being resolved into anything else by the eye, however much it might be magnified.

Here one has, set out in unmistakable terms, the same kind of differences which appear in viewing celestial detail through telescopes of various aperture. What cannot be seen at all with a low aperture may be seen with higher ones under totally different aspects; while in each case the apparent sharpness and clarity of the image is somewhat extraordinary.

Further in Fig. 189*b* in using the resolving power of the objective of high numerical aperture, the image may be quite wrecked by a little carelessness in focussing, or by mismanagement of light, so that one would hardly know that the valve had markings other than those seen with the objectives of lower aperture, and under these circumstances added magnification would do more harm than good. In precisely the same way mismanagement of the illumination in Fig. 189*a* would cause the striæ to vanish and with *Navicula Lyra*, as with many other diatoms, the resolution into striæ is a thing which often depends entirely on careful lighting, and the detail flashes into distinctness or vanishes with a suddenness which is altogether surprising. For "lighting" read "atmosphere," and you have just the sort of conditions that exist in telescope vision.

With respect to magnifying powers what has already been said is sufficient to indicate that on the whole the lowest power which discloses to the eye the detail within the reach of the resolving power of the objective is the most satisfactory.

Every increase above this magnifies all the optical faults of the telescope and the atmospheric difficulties as well, beside decreasing the diameter of the emergent pencil which enters the eye, and thereby causing serious loss of acuity. For the eye like any other optical instrument loses resolving power with decrease of effective aperture, and, besides, a very narrow beam entering it is subject to the interference of entoptic defects, such as floating motes and the like, to a serious extent.

Figure 190 shows from Cobb's experiments (Am. Jour. of Physiol, **35,** 335) the effect of reduction of ocular aperture upon acuity. The curve shows very plainly that for emergent pencils below a millimeter (1/25 inch) in diameter, visual acuity falls off almost in direct proportion to the decreasing aperture. Below this figure there can be only incidental gains, such as may be due to opening up double stars and simultaneously so diminishing the general illumination as to render the margins of the star discs a little less conspicuous.

An emergent pencil of this diameter is not quite sufficient for the average eye to utilize fully the available resolving power and some excess of magnification even though it actually diminishes visual acuity materially, may be of some service.

Increased acuity will of course be gained for the same magnification in using an objective of greater diameter, to say nothing

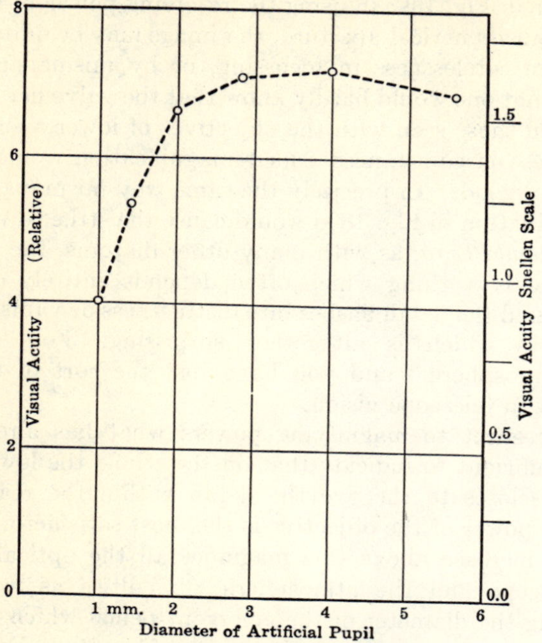

FIG. 190.—Resolving Power of the Eye.

of increased resolving power, at the cost, of course, of relatively greater atmospheric troubles.

To come down to figures as to the resolving power of the eye, often repeated experiments have shown that two points offering strong contrast with the background can be noted as separate by the normal eye when at an angular separation of about 3′ of arc. People, as we have seen, differ considerably in acuity so that now and then individuals will considerably better this figure, while others, far less keen sighted, may require a separation of 4′ or even 5′.

The pair of double stars ϵ_1, ϵ_2, Lyræ, separated by 3′ 27″ mags. nearly 4 and 5 respectively, can be seen as separate by those of fairly keen vision, while Mizar and Alcor, 11′ apart, seem thrown wide to nearly every one. On the other hand the writer has never known anybody who could separate the two components of Asterope of the Pleiades, distant a scant 2½′ but of mags. 6.5 and 7.0 only, while Pleione and Atlas, distance about 5¼′, mags. 6.5 and 4, are very easy.

Assuming for liberality that the separation constant is in the neighborhood of 5′ one can readily estimate the magnification that for any telescope will take full advantage of its resolving power. As we have already seen this resolving power is practically $\frac{4.''56}{A}$ for equal stars moderately bright. An objective of 4.56 inches aperture has a resolving constant of 1″ and to develop this should take a magnification of say 300, about 65 to the inch of aperture, requiring a focal length of ocular about 0.20 to 0.25 inch for telescopes of normal relative aperture, and pushing the emergent pencil down to little more than 0.02 inch,—rather further than is physiologically desirable. Except for these extreme stunts of separation, half to two thirds this power is preferable and conditions under which one can advantageously go above this limit are very rare indeed.

A thoroughly good objective or mirror will stand quite 100 magnification to the inch without, as the microscopist would say, "breaking down the image," but in at least nine cases out of ten the result will be decidedly unsatisfactory.

As the relative aperture of the instrument increases, other things being equal, one is driven to oculars of shorter and shorter focus to obtain the same magnification and soon gets into trouble. Very few oculars below 0.20 inch in focus are made, and such are rarely advisable, although occasionally in use down to 0.15 inch or thereabouts. The usual F/15 aperture is a figure quite probably as much due to the undesirability of extremely short focus oculars as to the easier corrections of the objective.

In the actual practice of experienced observers the indications of theory are well borne out. Data of the habits of many observers of double stars are of record and the accomplished veteran editor of *The Observatory*, Mr. T. Lewis, took the trouble in one of his admirable papers on "Double Star Astronomy" (Obs. **36,** 426) to tabulate from the original sources the practice of a large group of experts. The general result was to show the habitual use with telescopes of moderate size of powers around 50 per inch of aperture, now and then on special occasions raised to the neighborhood of 70 per inch.

But the data showed unequivocally just what has been already indicated, that large apertures, suffering severely as they generally do from turbulence of the air, will not ordinarily stand their due proportion of magnification. With the refractors of 24 inches aperture and upwards the records show that even in this double star work, where, if anywhere, high power counts, the general practice ran in the vicinity of 30 per inch of aperture.

Analyzing the data more completely in this respect Mr. Lewis found that the best practise of the skilled observers studied was approximately represented by the empirical equation

$$m = 140\sqrt{A}$$

Of course the actual figures must vary with the conditions of location and the general quality of the seeing, as well as the work in hand. For other than double star work the tendency will be generally toward lower powers. The details which depend on shade perception rather than visual acuity are usually hurt rather than helped when magnified beyond the point at which they are fairly resolved, quite as in the case of the microscope.

Now and then they may be made more distinct by the judicious use of shade glasses. Quite apart from the matter of the high powers which can advantageously be used on a telescope, one must for certain purposes consider the lowest powers which are fairly applicable. This question really turns on the largest utilizable emergent pencil from the eye piece. It used to be commonly stated that ⅛ inch for the emergent pencil was about a working maximum, leading to a magnification of 8 per inch of aperture of the objective. This in view of our present knowledge of the eye and its properties is too low an estimate of pupillary aperture. It is a fact which has been well known for more than a decade that in faint light, when the eye has become adapted to its situation, the pupil opens up to two or three times this diameter and there is no doubt that a fifth or a fourth of an inch aperture can be well utilized, provided the eye is properly dark-adapted. For scrutinizing faint objects, comet sweeping and the like, one should therefore have one ocular of very wide field and magnifying power of 4 or 5 per inch of aperture, the main point being to secure a field as wide is practicable. One may use for such purposes either a very wide field Huygenian, or, if cross-wires are to be used, a Kellner form. Fifty degrees of field is perfectly practicable with either. As regards the rest of the eyepiece equipment the observer may well suit his own convenience and resources. Usually one ocular of about half the maximum power provided will be found extremely convenient and perhaps oftener used than either the high or low power. Oculars of intermediate power and adapted for various purposes will generally find their way into any telescopic equipment. And as a last word do not expect to improve bad conditions by magnifying. If the seeing is bad with a low power, cap the telescope and await a better opportunity.

APPENDIX

WORK FOR THE TELESCOPE

To make at first hand the acquaintance of the celestial bodies is, in and of itself, worth the while, as leading the mind to a new sense of ultimate values. To tell the truth the modern man on the whole knows the Heavens less intimately than did his ancestors. He glances at his wrist-watch to learn the hour and at the almanac to identify the day. The rising and setting of the constellations, the wandering of the planets among the stars, the seasonal shifting of the sun's path—all these are a sealed book to him, and the intricate mysteries that lie in the background are quite unsuspected.

The telescope is the lifter of the cosmic veil, and even for merely disclosing the spectacular is a source of far-reaching enlightenment. But for the serious student it offers opportunities for the genuine advancement of human knowledge that are hard to underestimate. It is true that the great modern observatories can gather information on a scale that staggers the private investigator. But in this matter fortune favors the pertinacious, and the observer who settles to a line of deliberate investigation and patiently follows it is likely to find his reward. There is so much within the reach of powerful instruments only, that these are in the main turned to their own particular spheres of usefulness.

For modest equipment there is still plenty of work to do. The study of variable stars offers a vast field for exploration, most fruitful perhaps with respect to the irregular and long-period changes of which our own Sun offers an example. Even in solar study there are transient phenomena of sudden eruptions and of swift changes that escape the eye of the spectroheliograph, and admirable work can be done, and has been done, with small telescopes in studying the spectra of sun spots

Temporary stars visible to the naked eye or to the smallest instruments turn up every few years and their discovery has usually fallen to the lot of the somewhat rare astronomer, professional or amateur, who knows the field of stars as he knows the alphabet. The last three important novæ fell to the amateurs—two to the same man. Comets are to be had for the seeking by the persistent observer with an instrument of fair light-grasp and field; one distinguished amateur found a pair within a few days, acting on the theory that small comets are really common and should be looked for—most easily by one who knows his nebulæ, it should be added.

And within our small planetary system lies labor sufficient for generations. We know little even about the superficial characters of the planets, still less about their real physical condition. We are not even sure about the rotation periods of Venus and Neptune. The clue to many of the mysteries requires eternal vigilance rather than powerful equipment, for the appearance of temporary changes may tell the whole story. The old generation of astronomers who believed in the complete inviolability of celestial order has been for the most part gathered to its fathers, and we now realize that change is the law of the universe. Within the solar system there are planetary surfaces to be watched, asteroids to be scanned for variability or change of it, meteor swarms to be correlated with their sources, occultations to be minutely examined, and when one runs short of these, our nearest neighbor the Moon offers a wild and physically unknown country for exploration. It is suspected with good reason of dynamic changes, to say nothing of the possible last remnants of organic life.

Much of this work is well within the useful range of instruments of three to six inches aperture. The strategy of successful investigation is in turning attention upon those things which are within the scope of one's equipment, and selecting those which give promise of yielding to a well directed attack. And to this end efforts correlated with those of others are earnestly to be advised. It is hard to say too much of the usefulness of directed energies like those of the Variable Star Association and similar bodies. They not only organize activities to an important common end, but strengthen the morale of the individual observer.

INDEX

E

F

T

V

W

Z

A CATALOGUE OF SELECTED DOVER BOOKS IN ALL FIELDS OF INTEREST

A CATALOGUE OF SELECTED DOVER BOOKS IN ALL FIELDS OF INTEREST

THE AMERICAN SENATOR, Anthony Trollope. Little known, long unavailable Trollope novel on a grand scale. Here are humorous comment on American vs. English culture, and stunning portrayal of a heroine/villainess. Superb evocation of Victorian village life. 561pp. 5⅜ x 8½.
23801-6 Pa. $6.00

WAS IT MURDER? James Hilton. The author of *Lost Horizon* and *Goodbye, Mr. Chips* wrote one detective novel (under a pen-name) which was quickly forgotten and virtually lost, even at the height of Hilton's fame. This edition brings it back—a finely crafted public school puzzle resplendent with Hilton's stylish atmosphere. A thoroughly English thriller by the creator of Shangri-la. 252pp. 5⅜ x 8. (Available in U.S. only)
23774-5 Pa. $3.00

CENTRAL PARK: A PHOTOGRAPHIC GUIDE, Victor Laredo and Henry Hope Reed. 121 superb photographs show dramatic views of Central Park: Bethesda Fountain, Cleopatra's Needle, Sheep Meadow, the Blockhouse, plus people engaged in many park activities: ice skating, bike riding, etc. Captions by former Curator of Central Park, Henry Hope Reed, provide historical view, changes, etc. Also photos of N.Y. landmarks on park's periphery. 96pp. 8½ x 11. 23750-8 Pa. $4.50

NANTUCKET IN THE NINETEENTH CENTURY, Clay Lancaster. 180 rare photographs, stereographs, maps, drawings and floor plans recreate unique American island society. Authentic scenes of shipwreck, lighthouses, streets, homes are arranged in geographic sequence to provide walking-tour guide to old Nantucket existing today. Introduction, captions. 160pp. 8⅞ x 11¾. 23747-8 Pa. $6.95

STONE AND MAN: A PHOTOGRAPHIC EXPLORATION, Andreas Feininger. 106 photographs by *Life* photographer Feininger portray man's deep passion for stone through the ages. Stonehenge-like megaliths, fortified towns, sculpted marble and crumbling tenements show textures, beauties, fascination. 128pp. 9¼ x 10¾. 23756-7 Pa. $5.95

CIRCLES, A MATHEMATICAL VIEW, D. Pedoe. Fundamental aspects of college geometry, non-Euclidean geometry, and other branches of mathematics: representing circle by point. Poincare model, isoperimetric property, etc. Stimulating recreational reading. 66 figures. 96pp. 5⅝ x 8¼.
63698-4 Pa. $2.75

THE DISCOVERY OF NEPTUNE, Morton Grosser. Dramatic scientific history of the investigations leading up to the actual discovery of the eighth planet of our solar system. Lucid, well-researched book by well-known historian of science. 172pp. 5⅜ x 8½. 23726-5 Pa. $3.00

THE DEVIL'S DICTIONARY. Ambrose Bierce. Barbed, bitter, brilliant witticisms in the form of a dictionary. Best, most ferocious satire America has produced. 145pp. 5⅜ x 8½. 20487-1 Pa. $1.75

GEOMETRY, RELATIVITY AND THE FOURTH DIMENSION, Rudolf Rucker. Exposition of fourth dimension, means of visualization, concepts of relativity as Flatland characters continue adventures. Popular, easily followed yet accurate, profound. 141 illustrations. 133pp. 5⅜ x 8½.
23400-2 Pa. $2.75

THE ORIGIN OF LIFE, A. I. Oparin. Modern classic in biochemistry, the first rigorous examination of possible evolution of life from nitrocarbon compounds. Non-technical, easily followed. Total of 295pp. 5⅜ x 8½.
60213-3 Pa. $4.00

PLANETS, STARS AND GALAXIES, A. E. Fanning. Comprehensive introductory survey: the sun, solar system, stars, galaxies, universe, cosmology; quasars, radio stars, etc. 24pp. of photographs. 189pp. 5⅜ x 8½. (Available in U.S. only) 21680-2 Pa. $3.00

THE THIRTEEN BOOKS OF EUCLID'S ELEMENTS, translated with introduction and commentary by Sir Thomas L. Heath. Definitive edition. Textual and linguistic notes, mathematical analysis, 2500 years of critical commentary. Do not confuse with abridged school editions. Total of 1414pp. 5⅜ x 8½. 60088-2, 60089-0, 60090-4 Pa., Three-vol. set $18.50

DIALOGUES CONCERNING TWO NEW SCIENCES, Galileo Galilei. Encompassing 30 years of experiment and thought, these dialogues deal with geometric demonstrations of fracture of solid bodies, cohesion, leverage, speed of light and sound, pendulums, falling bodies, accelerated motion, etc. 300pp. 5⅜ x 8½. 60099-8 Pa. $4.00

Prices subject to change without notice.

Available at your book dealer or write for free catalogue to Dept. GI, Dover Publications, Inc., 180 Varick St., N.Y., N.Y. 10014. Dover publishes more than 175 books each year on science, elementary and advanced mathematics, biology, music, art, literary history, social sciences and other areas.